The Art of Exploration

*Lessons in Curiosity, Leadership
and Getting Things Done*

LEVISON WOOD

Illustrations by the author

HODDER &
STOUGHTON

First published in Great Britain in 2021 by Hodder & Stoughton
An Hachette UK company

2

Copyright © Levison Wood 2021

A CIP catalogue record for this title is available from the British Library

Hardback ISBN 9781529343021
Trade Paperback ISBN 9781529343038
eBook ISBN 9781529343045

Typeset in Bembo MT by Hewer Text UK Ltd, Edinburgh
Printed and bound in Great Britain by Clays Ltd, Elcograf S.p.A.

Hodder & Stoughton policy is to use papers that are natural, renewable
and recyclable products and made from wood grown in sustainable forests.
The logging and manufacturing processes are expected to conform
to the environmental regulations of the country of origin.

Hodder & Stoughton Ltd
Carmelite House
50 Victoria Embankment
London EC4Y 0DZ

www.hodder.co.uk

All that is gold does not glitter,
Not all those who wander are lost
  – J.R.R. Tolkien

For my brother, Peter

Contents

Introduction

Why am I telling you all these examples of exploration and adventure? Because we are all explorers in life, whichever path we follow – Nansen, *Spirit of Adventure*, 1926

I had never been more scared in my life.

'Jump! Just jump!'

I was a child. A terrified seven-year-old child. Since my earliest thoughts I had dreamed of adventure, but now that I looked down at the swirling waters beneath a sheer cliff face, I was beginning to realise that I was not the explorer I had imagined myself to be.

'Jump!'

Right there, in that moment of childhood, I had reached a fork in life's road. I could never have known it then, but to step back from the cliff's edge would have sent me on a very different journey to the life that I live today. If I had backed away from that cliff, I might never have gone on to jump out of planes with the Parachute Regiment or walk through war zones with nothing but my camera. I may not have experienced the differences between the sodden heat of the jungle and the baking sun of the desert; breathed in the smells of Himalayan meadows, or felt the ground move as a herd of elephants passed by.

Would I have ever shared smiles and conversations on the world's frontlines, or in remote villages forgotten by time? I don't believe I would have done.

There is a whole world for us to explore, but it will always be out of reach until we take the first step.

Until we jump.

Fortunately for me, I had my father with me on that cliff top, if you can even call it that. It seemed miles high at the time, and there was no way that I could have mustered the courage to jump into the cold waters below unless my dad had asked, 'Do you trust me?'

I did, and even though I thought we were probably jumping to our deaths, I followed him over the edge, and to adventure. From that point on in my life, I could not be held back. Now, in this book, I want to do for you what my father did for me.

I want to support, guide and encourage you to think like an explorer.

By virtue of the fact you have picked up this book, you have shown that you are a curious individual, and curiosity is what drives exploration. It is the essence of humanity. Our ability to ask questions, to learn and to share that knowledge is what has made our species the most successful on the planet. Exploration is hardwired into our DNA, so if you feel like you are sometimes consumed by an urge to discover new things, to travel or to explore, then you are not alone.

Because it is who we are.

Too often we can feel like passengers in our own lives. This is because we have surrendered our curiosity to other people. We let them make our decisions, and tell us what shape our lives will take. We let fear and mistrust guide us; we become scared of change. We allow ourselves to fall into a rut – our comfort zones – and fail to appreciate new things. We forget that when you retrieve your own curiosity, you can begin to take back control of your life and start sitting in the driving seat .

I can tell you from experience that being a passenger in a car that is out of control is not a good place to be. I was very lucky to survive when the taxi I was travelling in fell hundreds of feet from a Himalayan mountain road. I had no right to survive that accident, and yet here I am. I have been very lucky in many ways in life. I survived the war in Afghanistan too, and some close calls with crocodiles and elephants in Africa. Somehow, I narrowly missed stepping on several bombs in Iraq and Syria. For that, I am very grateful.

I was fortunate to grow up in a stable democracy and I couldn't have asked to be born to better parents, who believed in the value of education and gave me the gift of curiosity. But even with the privilege of a very happy childhood, I had no idea that my dreams of exploring the world would ever become true. We were not wealthy, and I assumed that the only people who could have those kinds of adventures were the very rich. I never imagined that it would be possible to make my living by following my dreams of travel.

Some people are born with the travel bug, others are later bitten by it, but the need to visit far-flung places is not a criteria for reading this book. It is about the importance of allowing yourself to be curious, and to become a student of life.

I'm not trying to convince you to become a professional, full-time explorer, and that's not because I don't want the competition! Part of the explorer's mindset is to be realistic, and even though what appear at first to be obstacles can often provide the answer, there are certain things in our life that restrict our ability to do as we like. If your dreams are to explore full-time, then you already have the explorer's mindset, and I hope you will find in these pages the tools to refine those raw ideas into an art form.

For everyone else, no matter your age or profession, I believe wholeheartedly that these lessons from my travels can help you to fulfill your potential for living a happy life, regardless of your circumstances.

You might feel that you keep getting knocked by the world and the people in it, and you need help to get back on your feet. Or you may be winning at life and constantly looking for ways to sharpen your mental blade. Whatever position you're in, you can always do better. We can always learn. We can always improve. If you have waited sixty years to explore, but start today, that is not failure. The start point will be different for everyone, but the destination will be the same – a happier, more fulfilled life.

I have always tried to learn from my own mistakes and from those of others. I have also learned from the successes of strangers, taking inspiration from people who have done things that I want to do, applying those lessons to create the life of my choice. Life is all about making choices. Being human is all about making choices; but we can only make those choices when we recognise first that we have them by taking ownership of our lives and taking the relevant action. There is much in the world beyond our control, but that it is no reason to surrender the huge amount of power that we *do* have over our own lives.

Affirmative action is a term we used in the military to describe a situation when you need to do *something*, even if you don't know what, in order to progress and avoid getting stuck in a negative pattern. Moving forward breeds success, and the more considered action we take, the more successful we will be in any line of work.

This book sets out to arrange the ways that I have learned to act and think. They are the guiding principles that have given me the best opportunities in my vocation, which is so intertwined with my life that it is impossible to separate the two; nor would I want to, because I love what I do, and I do it because I can't imagine doing anything else. It is my purpose. I used to shy away from the term 'explorer', as it felt a bit old-fashioned and fusty, belonging to an era when men wore pith helmets and shot tigers. Now exploration takes on a new meaning and is far more inclusive. It is about documenting a moment in time, not for the sake of posterity, but for the here and now, so that we can all share in the dissemination of learning, improvement and collective knowledge.

A hundred and fifty years ago, vast swathes of the world remained unexplored. There were countless opportunities to undertake pioneering expeditions. Many rivers had yet to be navigated, almost no great mountains had been summited and the poles were thought to be an impossible dream. Today, you can simply look at your phone and scroll on Google Maps to see any point on our planet. There is no need to worry what will happen when you get there, either – Trip Advisor will fill in all the details. Almost no 'firsts' remain to be conquered. Instead, would-be explorers have taken to adapting the well-trodden paths with new speed records, unassisted or oxygen-free climbs, and eco-friendly adventures. Good for them, I say!

There remain plenty of reasons to travel, beyond verifying what the digital guidebooks say, and finding a new variation on an old route is a worthy undertaking in itself. It is better than not travelling at all, that's for sure. But fear not, there is still one unexplored journey to be had for each and every one of us, and that is in the quest for self-discovery – to live a meaningful life. This kind of exploration traverses more than the physical realm, but the treasures are perhaps even more rewarding.

There are risks, of course, and no GPS can tell you how to live a good life. Therein lies the challenge. In this modern age of chaotic technological advancement, even Google cannot find out who you truly are, and the signposts are perhaps more obscured than ever. That is why it is vitally important to forge your own path and plot a course towards a life of your own making. This is what I have always tried to do and I hope this book will help you to do the same.

To that end, I have included lessons from some of the great explorers and figures in history, but also anecdotes from my own life – from childhood adventures and student wanderings to military scrapes and my encounters as a professional explorer. As you will see, I have never had all of the answers, but fortunately there have been people around me who helped out along the way, and now I want to pass their wisdom onto you. I have been fortunate to travel to more than a hundred countries, and wherever I have gone I have tried to be curious and learn lessons from the people around me. In that sense, exploration never ceases.

I have set out these eleven lessons from the road not as a comprehensive rule book, nor as an autobiographical summary of my own adventures, but simply as a way of showing how the art of exploration can benefit anyone.

This is a book for students, entrepreneurs, career-changers, armchair wanderers and veteran travellers alike. It is about curiosity – and that belongs to us all. It is part philosophy and part smart-thinking, with a smattering of guidance.

Life is a complicated matter and we all have our own path to follow and mission to achieve. For some people, winning comes easy. For others it may take a lifetime. But what is for sure is that we can all be explorers in our own right if only we adopt the right thought process. So forget routine; now is the time to embrace the unknown, step out of your comfort zone and open the gateway to the Art of Exploration.

All you have to do is jump.

Levison Wood
London, January 2021

Sir Richard Burton

I

Know Yourself

Knowing yourself is the beginning of all wisdom – Aristotle

Exploration starts at home

I have a rather unusual name, and a rather unusual career. That often leads people to assume that I grew up somewhere far away and exotic, and that my parents brought me up amongst the local wildlife on some far-flung frontier. They're often disappointed when they hear the truth of where I was born and raised.

Stoke.

Nothing much exciting happened in Stoke in the early 1980s. The coalmines and factories were all but closed, and there was widespread unemployment. People's ambitions were mostly limited to a ten-mile radius and a sense of the foreign was encountered only when the first curry houses started to open. Families were tight and the idea of going off on an adventure was complete anathema.

Not that I'm complaining. Despite having a typical and very average upbringing, Stoke-on-Trent was very good to me. My childhood memories are a bundled collection of passing images: my little brother in a pram being pushed by my Auntie Josie outside my grandparents' house; walking up a cobbled side street in Hanley city centre going shopping with my mum, and standing in a red soldier's uniform in the school play. They are all visions of home; a happy childhood in a safe place, far from exotic, a mundane vision of suburbia.

That said, there are a few moments I do recall that set me upon a path of discovery when I grew up; the first of which came when I was about six years old. It was morning assembly at my primary school, all the pupils sitting cross-legged in the hall, eagerly awaiting

Mr Adams to address us. 'Mr A' as he liked to be known, was our favourite teacher. Kind and funny, he was the only male teacher at the primary school, memorable for being tall and hairy. He let us play and never shouted; he didn't need to, we all did as we were told.

Anyway, we sat in silence, and I remember sitting in the second row to the front at the far left, looking towards the line of chairs where the other teachers, Mrs Rogers and Mrs Watts, were sitting. Mr Adams stood up.

It was story time.

Sometimes we were read a few pages from Enid Blyton's *The Famous Five*. Other times it was a biblical proverb regaling the adventures of some long-lost saint. But this morning Mr Adams had a treat for us. It was a real-life adventure – something that had happened to him personally. When Mr Adams was eight, he had sneaked out of his house with two friends, and walked across the fields in search of an elusive newt pond. Armed only with some jam jars and a net, the boys went off on a voyage of discovery for this mysterious Shangri-La of Stoke-on-Trent.

These were local fields, which I'd heard of myself. He travelled far, crossing over the brook, and climbing over a fence and through a small forest, until it was very late in the afternoon and he realised that he was lost. His friends ran away, abandoning Paul (as he was known to them). As the sun set, Paul followed a hedgerow towards a hill in the distance. At the top he looked out over the forest to the fields beyond, where he thought he could make out the familiar sight of a nearby town.

Looking down, he saw a path between the trees, but that led to a cliff, and beneath it was a gorge through which ran the railway track. He knew it was dangerous to go anywhere near the train tracks, and there were signs everywhere with a skull and crossbones – but he had no choice. He climbed down the little cliff, sending soil and rocks crumbling down the sandstone divide, until he found himself standing, alone, on the gravel embankment. Looking left then right, he stood there for five, maybe ten minutes, fearful that a train might come and crush him at any moment.

Eventually, he made the mad dash, hopping over the metal girders and the wooden beams until he was safe on the far side. From there, he was able to climb up the other bank and follow a path that led him to safety; through the woods, across the fields and towards the town, where he was able to walk home.

I'm sure there must have been some lofty moral to the story. All the other stories, proverbs and biblical extracts we'd been read had a moralistic tone, and this one would have been no different, even if it was real life. However, at the time, this little tale of misadventure left a lasting impression on me. I don't know why I remember it so well, because it wasn't particularly memorable or adventurous, especially when compared to the stories of my own dad or grandad, who had served in the war. But it was probably the first one I'd heard where the protagonist was not an immediate family member, and he got away with doing something naughty.

Either way, the memory stuck. I thought how wonderful it was to be a grown-up boy of eight, who could go off on an adventure and get lost, and be able to climb down cliffs and run over train tracks. It sounded like the height of exploration. I also remember how disappointed I was. I was already six, and I hadn't done anything so extreme or dangerous yet. I felt as if I had wasted my life so far, and promised myself that as soon as I was old enough, I would go off on a perilous adventure. I wanted to be just like him and have my own stories to tell.

But Mr Adams' story was only an idea. We all have those – dreams where we act out the life that we want, rather than accept the one that we lead – and the truth is that I did not run away from home on my own newt-seeking adventure, or risk life and limb to climb down cliffs and cross train tracks at the next opportunity, much to my parents' relief.

Perhaps my dad did recognise something in me, though. A chafing desire to be outside and explore the big wide world. Either way, a few months after I heard Mr Adams' story, my dad took me canoeing and camping for the first time.

It was the height of summer and we paddled down an ancient gorge; suddenly I felt as if I was exploring the world myself, and not hearing about it second-hand through other people's stories. I loved the feeling of floating down the fast-flowing river, my dad telling me to hold on tight – there was a sensation of danger, and it was thrilling.

After some time, we came to a pebble beach under the shade of a tree. My dad pulled us up onto the bank. 'Come on, son, it's time,' he said purposefully.

Time for what? I thought, as we got out of the canoe, but I didn't ask. I could feel we were on the edge of something important, and I didn't want to ruin it with questions.

'Follow me,' he said, and leaving the paddles behind, we scrambled along the water's edge, and up a rough track that ran parallel to the river. My dad helped me up the rocky outcrops, until we were standing on top of a cliff that seemed to tower over the whirling rapids below. I assumed that he'd brought me up here so that I could admire the view.

I was wrong.

'Jump,' he said, still holding my hand.

Jump? It seemed unfathomably high and I could barely bring myself to peer over the edge. Instead I looked at him and started to cry. There was no way I was going over that cliff. I wanted adventure, but it seemed obvious that to jump down there was to die.

'Son,' my dad said, with a patient stare, 'do you trust me?'

I was silent. I had a fear of heights and I didn't much like water, either. The combination terrified me. I could hear the water crashing below. In my mind, it was a mile to the bottom; in reality, it was probably only a few metres.

'Would I ever hurt you?' my dad asked benevolently.

Ashamed of my cowardice, I looked at my feet. 'No . . .'

'We'll jump together,' he said. 'Okay?'

So the decision was made for me. We'd die together, but better death than the shame of letting my father down.

I closed my eyes and hoped that my feet would grow roots into the stone. They did not, and my knees trembled as my dad slowly pulled us towards the edge.

'Jump!'

I held my breath and clenched my hand in his, as we stepped forwards and into the abyss.

The fall seemed to go on forever. My eyes remained closed as I held my breath, waiting for the impact. Everything was silent as I plunged down through the air, my fingers suddenly slipping free, and I could no longer feel the safety of my father's hands.

Whoosh . . . I hit the cold waters and felt my head sucked under the eddy. I wanted to let out a cry of shock as I opened my eyes and saw nothing but white foam whirling all around. I had no concept of up or down. I might have been pulled a mile downstream for all I knew, but I kept holding my breath, hoping for the best. Suddenly, I felt someone grab me from behind and pull me up. I turned around and let out a sigh of relief when I felt my dad close by. He lifted me clean out of the water and had a big smile on his face.

I wanted to cry again, but I didn't – I was safe! The overwhelming sensation changed instantly from terror to relief, which morphed into a visceral bodily pleasure. I found myself inexplicably happy, beaming with confidence and joy. I had done something I never thought I could do, and to top it off, my dad was proud of me!

What did I do next? I swam back across the river, of course, climbed to the top off the cliff and jumped off again, this time on my own, and without a moment's hesitation.

I had conquered my fear, and felt invincible.

The motto of my old high school is *Know Yourself*.

Painsley was a Roman Catholic comprehensive school which drew in a wide range of children from across the surrounding towns and villages due to its good reputation for the sciences, and a stern disciplinarian staff. Mr Tunney, the headmaster, would explain to all the new pupils the importance of knowing yourself.

'Knowing yourself is the beginning of all wisdom,' the head would say, quoting Aristotle, and we would stare back blankly. Everybody knows themselves already, surely? Why would we need to be reminded of that?

Perhaps the school motto was not intended for the children we were then, but for the adults we would grow up to be. It was a reminder that knowing oneself is the key to both success and happiness. Being self-aware means marrying your outward behaviour and actions with your internal ideals and values. Astute self-awareness is a good predictor of success in life; it gives people a clear idea of the opportunities that will suit them and those that won't.

Above all, this ability to see yourself as being separate from your environment can bring you happiness and contentment. As the Chinese philosopher Lao Tzu put it, 'Knowing others is intelligence; knowing yourself is true wisdom.'

Samsara

If I asked you to think of an explorer, His Holiness the Dalai Lama might not be the first person who came to mind, but he is a man who trekked over the Himalayas on foot, travelling at night and in disguise to escape persecution in his homeland. What's more, he has since travelled the world spreading the message of peace, reconciliation and compassion, whilst maintaining his very humble origins. He is a man of boundless energy, infectious enthusiasm and a childlike curiosity – all characteristics of a great explorer.

I was fortunate to meet and speak with the man himself, when I visited India and listened to one of his sermons on the subject of 'Samsara', a central tenet of Tibetan Buddhism. Samsara is a Sanskrit word that translates as 'wandering', but in this context it can also mean rebirth, and life after death – reincarnation. Its essence is a cyclic and circuitous change known in Buddhism as the karmic cycle. It is best understood as 'a cycle of aimless drifting, wandering and mundane existence'. The whole point of our spiritual quest, according to the Dalai Lama, is to be liberated from Samsara, through enlightenment. This can only be achieved through self-awareness, understanding and self-mastery.

For Buddhists, and many others who follow the ancient Eastern philosophies, the meaning of life is simple: to achieve a state of mind that is free from desire, temptation and greed, and to walk a middle path of balance and moderation.

Very few people achieve it, but with dedication and hard work, we are all capable of it. For me, the Art of Exploration is about discovering the world with purpose and intent, and an aspiration of betterment. The starting point for our liberation from aimless wandering is self-knowledge.

The importance of self-awareness

The year 2020 will be remembered as the Great Pause. As the world went into lockdown, billions of people were quarantined in their own homes for weeks and months on end due to the Coronavirus Pandemic. For thousands across the globe it was a time of tragedy, losing loved ones to this terrible disease. For many millions more it was financially ruinous, as jobs and businesses went under and the economy plummeted.

Of course, it affected many people in different ways, but despite its awful impact and terrible consequences, if we were to fathom some positives to come out of the whole mess, it would be fair to say that it brought many communities together, unified in purpose and a desire to help one another get through it. Perhaps even more importantly, it forced a great number of us to take stock and reflect on our lives in a way that we never have before. It certainly did for me.

It is quite remarkable what a few weeks of solitary confinement can do for you. Shakespeare wrote *King Lear* while quarantining from the plague and the playhouses were closed, and the famous seventeenth-century diarist Samuel Pepys documented the impact of the rampant 1665 bubonic plague in London:

> But, Lord! how sad a sight it is to see the streets empty of people, and very few upon the 'Change. Jealous of every door that one sees shut up, lest it should be the plague; and about us two shops in three, if not more, generally shut up.

Seclusion has also afforded a great many writers over the centuries the chance to come up with their finest masterpieces. Dante wrote *The Divine Comedy* whilst in exile, and Cervantes came up with *Don Quixote* whilst behind bars. Dostoyevsky too was inspired to write two of his finest works after spending months in jail, and let's not forget Nelson Mandela, and a whole host of other political leaders. It seems that having one's liberty removed temporarily, if looked at with a positive mindset, can enable you to focus on things that really matter; and that begins with self-reflection and an understanding of oneself.

I am certain that, in time, we will look back at 2020 as being a year of catalytic change in many ways – politically, economically, and socially. We have lived through historic times, and I believe that much of the change to come will be driven by people who have used the time wisely, thinking about what they can do to improve themselves. There are the obvious things that many of us aspire to, such as reading books, getting fit, learning an instrument or a new language, or perhaps taking up a new hobby, whether that's origami or baking. But as well as the 'easy wins', I'm also referring to deep, fundamental changes in how we go about our daily lives, committing ourselves to a new regime of betterment in how we treat ourselves and others – and all this begins with understanding.

I've heard more than one person say that – if not for the losses that so many people have suffered – they were almost glad the pandemic happened, because it gave them the first chance of a break in their lives for decades. If it takes a coronavirus to give you the time to do what you want in life, then that would suggest a strong case for examining how you live; and you can only do that if you are self-aware. We can make big changes at any point in our lives, but we often get distracted because of work, relationships and other external factors, and it is easy to ignore what is happening deep inside of us. As well as taking care of our physical health, we need to carve out time to work on our self-awareness.

There are few times in our lives when we are forced to be still and make peace with our decisions, because we have no choice but to do otherwise. For me as an explorer and professional traveller, the year 2020 was the first time in over a decade that I'd spent more than a couple of months in one place, so it was certainly a big change from the norm.

At first I remember feeling trapped and a bit claustrophobic, stuck in London when all I wanted was to enjoy the freedom of the road. All my trips got cancelled and, like many freelancers, I lost an entire year's worth of wages. It was made harder by the fact that I ended up breaking an ankle, which forced me to stay put, even if there was a temptation at times to escape. It would have been simple to just sit in front of the TV and do nothing, but, instead of moping, I decided to try and use the time wisely and write down some of the lessons I have learned from travel, and in doing so reflect on what I could do better.

It reminded me of the last time I was forced to stay still and take stock in the summer of 2015. It was under slightly different circumstances, but with a rather similar outcome. I remember the date well – it was 19 August. I was on an expedition in the Himalayas, when the taxi I was travelling in took a tumble off the edge of a cliff in the dead of night, and I was plunged into a jungle ravine. Somehow I survived and escaped that time with only a broken arm and a few smashed ribs, but I ended up having to halt the journey for fifty days while I recovered from the accident. I'm sure a close brush with death is enough to give anyone pause for thought, and it forced me to re-evaluate things and ask myself a few questions about my own life, who I was, and where I was going.

Quite often we can get on a path to reach a 'destination', but neglect to re-evaluate the next stage once we get there: we simply keep going in the same direction without thinking. This can happen in business, relationships or in any part of our life. How often do you sit back and think – *really think* – about whether or not the path you are on is the right one? Maybe it was the right choice for you a few years ago, but life changes us, and what works for the person you were *then*, may not be right for the person you are *now*.

Wilful ignorance

Like many young men, I spent much of my teenage years unsure of who I was and where I would fit into the world. I did have a strong idea of what I wasn't about – I disliked routine and knew that a 9–5 office job was never going to satisfy me – but I found it hard to identify something positive that I could sink my teeth into. Back then, I thought that only the really wealthy could live the kind of life that I wanted.

I was fiercely independent, but I was also driven intensely by social approval in ways that I wasn't even aware. In fact, I think a lot of what I thought of as my 'independence' at the time was driven by a need for other people to see me that way. In my desire to be seen as making my own choices, I was making those choices because I was worried what people would think of me!

I was also hugely competitive. I couldn't bear the thought of other people seeing things in the world that I wouldn't see, having experiences that I wouldn't have or being able to do certain things better than me. I wanted to experience *everything*, all at once, and I was consumed by a constant fear of missing out. In my early years, it was fear that drove me. I was scared of being denied experiences, and ironically this led to an almost self-fulfilling prophecy as you cannot be truly grateful and appreciate the experiences that you *do have*, if your mind is whirling at 1000 miles an hour, trying to push you on to the next one. I have definitely been guilty of tick-in-the-box tourism.

As a young man, I started to drift. There are a number of structures that you can pin your efforts to at that age – sports teams, academic exams, and the like – which give you a sense of progress and purpose. But as time goes by, these gradually fall away, like scaffolding coming down to reveal a half-completed building. By my early twenties, I was starting to realise that I'd been going through life accumulating achievements and experiences, clocking up new pins on my mental map of the world in all the countries I'd visited, without ever really asking myself what was the underlying purpose of all this exploring. It took me a long time to figure that out.

Having an explorer's mindset is not about visiting new or challenging places for the sake of visiting them. You have to ask yourself not only where or what you want to explore, but *why*. If you are facing challenges in your life, a bit of travel might seem like one way of getting a sense of perspective on these (more on this later), but it might not, because *you* are *you*, no matter which continent you stand in. Either way, what's important is not where you go or what you do, but the mindset that you travel with. It's important to want to learn about the world around you, but the answers will never fit into place until you understand the person asking the questions.

The First Adventure

Homer's Ancient Greek epic poem, the Odyssey, is perhaps the earliest travel story in Western literature. It tells the adventures of Odysseus, the king of a small Greek island called Ithaca. Odysseus leaves his wife, Penelope, and their baby son at home and sets sail to fight in the

Trojan war. The war drags on for a decade, but that isn't the end of his travels. In his mission to get home, the fraught king encounters all sorts of obstacles. The journey is filled with fantastic tales of passionate affairs, a trip to the underworld and encounters with horrific monsters and demons.

Odysseus is the hero of the story, but he is deeply flawed, and brings a lot of the tale's misfortune upon himself. His hubris after successfully tricking the Cyclops leads to him being cursed by the sea-god Poseidon, and he must wander the Mediterranean for ten years before he can return home to his family. His lack of self-awareness sets him at odds with his own compatriots – the Ithacans. Odysseus is warned not to eat the sun god Helios's sacred cows, but his men ignore this and are shipwrecked as a result, with all but Odysseus dying. Then, when he returns to Ithaca, the man who was presumed dead murders his wife's suitors in a cruel rampage, torturing some, and sparing none. Odysseus is blamed for the deaths of two generations of Ithacans – all because of his egotistical blunderings, and a lack of self-awareness.

This story is foundational to how we imagine adventure, at least in the West. It has underpinned our collective imagination for nearly three thousand years, and the word 'odyssey' – derived from Odysseus – has become synonymous with long journeys. The Ancient Greeks, along with Aristotle, were the first Westerners to explore ideas of self-knowledge in a meaningful way. This is reflected in literature from the period, and is the foundation of what we now call 'philosophy'.

Discover your inner explorer

When we start to ask questions of ourselves, inevitably we get drawn towards the notion of 'purpose'. Why am I here? Where am I going in life? What do I want to achieve? It is important to ask these questions, because without them we drift aimlessly. We might have a good idea of what we need to do to keep going from one day to the next, but without being able to tie this into some higher overarching goal or set of ambitions, we quickly become disillusioned.

We don't all want the same things, and we (quite rightly) have our own personal interpretations of success. That makes it even more difficult when it comes to trying to distil lessons from other people's lives. No one should presume to tell anyone else what they should be striving for, because that is a very personal matter, but for me the successful explorer's mindset begins with a general principle, summarised aptly by Ernest Hemmingway: 'There is nothing noble in being superior to your fellow man; true nobility is being superior to your former self.'

I echo Hemingway's words. I truly believe that success is simply doing better than you were before; we all have a duty to understand ourselves better, and to strive constantly to improve ourselves, because life is about learning lessons. Classical philosophers from Aristotle to the Buddha and Lao Tzu have identified knowing ourselves as the starting point for a meaningful existence, but what does self-knowledge mean? What does it take to know oneself?

To understand our goals, we need to understand our flaws; to understand our flaws, we need to understand ourselves. Our histories (ancestral and personal), our identity, our upbringing, our desires and our values. To begin the journey of self-improvement, we must understand and grapple with the various factors, mistakes, hopes, dreams, strengths and weaknesses that define us as individuals.

This can be very uncomfortable and difficult work, but if we are truly to embrace the mindset of an explorer – someone who can think for themselves, be free of judgement, and aspire to humility and gratitude – then we need to accept that there is something we *can* do to help ourselves.

First, however, we need to understand a little bit more about our own drivers and motivations. As T.S. Eliot famously wrote: 'We shall not cease from exploration / And the end of all our exploring / Will be to arrive where we started / And know the place for the first time.'

It is impossible for us to know our limits and the extent of our powers without testing ourselves, pushing a bit further than we thought possible and delving a little deeper into the unknown, and we don't know who we will be when we return after our own personal odyssey.

Knowing yourself is important no matter who you are, and what you do. Navigating through life has always been difficult, but now we are under constant bombardment from our screens and phones telling us who we should be, and what we should think. We work long hours and are often tired, so it can be very tempting to accept one of these 'one size fits all' personalities that social media tries to hand us.

The problem is that, as humans, there is no one size fits all. We are all unique, and we need to make time and space to ask ourselves the question: what is truly meaningful to us? Without knowing the answer to this question, we cannot be effective, caring, understanding, calm or purposeful in what we do. In the hustle and bustle of modern-day living, we can get swept up in routine, work and family, without spending time to consider who we want to be and where we really want to go.

To live a fulfilled, exciting and happy life, I believe we have to create our own philosophy, whether conscious of it or not.

Sir Richard Burton

One of my personal role models and childhood heroes is the Victorian explorer Sir Richard Burton. Long before I was old enough to be let loose on my own, I read all about his adventures in faraway lands, using my imagination to fulfil my lust for adventure. He was a man of many talents, earning his keep as a soldier, diplomat, spy, translator and author. Far from ordinary, Burton was a troublemaker in his youth, known for clouting a teacher over the head with his violin. He was thrown out of Oxford University for violating dozens of rules (including keeping a pet bear), and left in an enormous huff, flattening the college flower beds on his way out with his horse and carriage.

As he grew up, Burton learned how to behave, but retained a rebellious streak. He hated what he called the 'slavery of civilisation' and rejoiced in shocking polite society. A young vicar once asked if it was true that he had killed a man in the Arabian desert. 'Sir,' Burton replied coolly, 'I'm proud to say that I have committed every sin in the Decalogue.'

But Burton was not merely a hot-headed rogue. He had a sharp intellect and a healthy approach to improving himself; possessing the ability to work hard at subjects that interested him, or that he felt served him. *He was a polyglot, with a mastery of multiple European and Asian languages, as well as multiple dialects of Arabic.*

Free-spirited by nature, he would remain outspoken throughout his life and his adventures included a contested and controversial journey to search for the source of the Nile. His weather-worn complexion, dark eyes and language skills made him a perfect spy for the East India Company, but perhaps the greatest example of his guile and guts came when he smuggled himself across the deserts of Arabia and into Mecca, which was and is completely forbidden to non-Muslims. Not one to do things by halves, Burton had completed his disguise by getting circumcised.

The explorer once wrote, 'Of the gladdest moments in human life, methinks is the departure upon a distant journey to unknown lands,' to which I can certainly relate. There is no better feeling than the immense anticipation that comes with embarking on a trip of any kind. Burton found life in the desert and in the wilderness – and the risks of getting caught – utterly exhilarating. The idea of this endless inquisitiveness, and of being a brave and fearless risk taker, is what appealed to me as a boy. Burton was forever seeking out new experiences and continually defied expectation. I admire that he accepted his exploratory instincts and embraced his own insatiable curiosity.

In this respect, he is the quintessential Victorian explorer, determined to make his name in the annals of history, and this is often the focus when it comes to Burton. Beneath all of this, though, there is something far more nuanced and that is his enlightened attitude to self-awareness. It is summed up well in the quote, 'Do what thy manhood bids thee do, from none but self expect applause.' It is important to understand that when he wrote this, the word manhood meant 'the condition of being human'.

Burton made great pains and efforts to improve his understanding of the self. He was convinced that the best measure of our success was against ourselves, and that the noblest claim was to adhere to one's own moral compass. He believed that without this understanding, you were no better off than a slave.

Finding your quest

I believe that it was Burton's capacity for introspection and reflection that set him apart from so many others, and that this trait is something worth emulating. Giving ongoing attention to one's internal state and one's inner life is what will make that life rewarding. When it comes to travelling, are we doing it to help answer questions about the world and our place in it? Or, without that introspection, are our travels nothing but vanity – a slide show to let everyone know how adventurous and fortunate we are?

There's a great meme of a couple lounging on a beach with cocktails in hand, and palm trees in the background. The caption reads: 'Oh no, we're still us!' It's meant to be a caution for travellers who think that the simple act of moving from place to place is the remedy for everything.

If your problems exist inside your head and heart, then they will be with you whether you're in the Taj Mahal or a Tesco carpark. You can't outrun thoughts and feelings, but travel can help you process and place them, so long as you approach it with the purpose of self-discovery. Only with an open mind can we learn more about our inner purpose – what we want, what we think and what kind of world we want to leave behind us.

Of course, people travel for all sorts of reasons. The truth is that very few will go to booking.com and put 'somewhere I can find myself and the meaning of life' into the search box. Self-discovery is often a by-product of another venture, at least at first. That is why having an open mind is so important. So long as we put ourselves in positions to learn, and don't fight the revelations when they come, we will grow by osmosis. If you travel to discover new culture, who you are as a person will change. If you travel to learn a language, that ability to connect more deeply to a whole new population will also change you. If you travel for physical adventure, that challenge will change you, too.

There is no right or wrong about the reasons for travel. Even if the sole aim of one's journey is to relax, then that too is sufficient, so long as it gives time for thoughtfulness. Rushing around from sight to sight, ticking off temples and flitting from Instagrammable beach to five-star resort may boost your ego, but ultimately it does little more than send you home more exhausted than before you left. Living this way is seeking other people's approval, not your own.

Honestly, I can only imagine how much more shallow my early travel experiences would have been if social media was around when I was a younger man. Yet I think a spell of tick-in-the-box tourism is natural for most of us. If you grow up in a consumer society, then it is silly to think that the mentality of 'gotta have it all' would not stretch to travel. The key is to recognise when that is happening, because it's a behaviour that you can relapse into at any time.

Did you book a trip to impress someone? Are you spending more time looking at your phone than the sights you came to see? Does your enjoyment of the trip diminish because not enough people comment and like your social media posts about it? We can all slip into these negative loops at any time, unless we are self-aware enough to know that they *will* happen. Don't feel bad when they do. Give yourself a little pat on the back for noticing, and refocus on what is really important; even if you did book a trip for the wrong reasons, that doesn't mean you can't use it as a chance for personal growth while you're there!

You would not be the first person who travels to try and escape their problems. Travel provides a method of escape from the mundane, normal and tragic. Going on a journey has long been seen as the surest way of getting over a broken heart, and everyone knows a good love story that involves the protagonist fleeing his or her breakdown and going off to a faraway land.

Travel is escapism in its rawest form. It is a way of getting away from life's distractions, which in itself can be both good and bad. They say that running away from your problems solves nothing, and whilst this is true, I don't think there's anything wrong with taking a break from your problems, so long as you understand that they will still be there when you get back. When I broke my arm, I knew that at some point I would have a lot of hard work to do in physio, but was I jumping out of bed to do press-ups on day one? Absolutely not! It is okay to rest broken bones, broken hearts and battered minds, so long as we use that rest to prepare ourselves for the hard work that lies ahead.

Sometimes you need to take yourself out of the equation for a while in order to gather your thoughts, gain a new perspective and look at the problem from an alternative angle. You never know, you might have that lightbulb moment that inspires you to make the necessary changes. You might find a random stranger to share your problems with on a train, and they suggest some useful advice; or you might need to go somewhere with a different culture, which offers up a novel outlook. Even if you don't, a change of scenery and a long walk will at least give you the time and space to process what you have left behind. Sometimes you simply need to go away in order to return.

Travel has gifted me endless experiences; it has given me shocks, surprises and tragedies. I have witnessed suffering beyond comprehension, seen poverty beyond words and uncovered awful abuse. I live in a very safe and comfortable corner of the world, and I could stay here and avoid all of the terrible things I have seen, and yet I still travel for the sake of it, and I seek out the hardest places in the world. It has enabled me to formulate my own viewpoints based on real-life experiences, and not merely theories or speculation. When you travel, you realise that all those experiences, both good and bad, teach you a lesson.

Gustav Flaubert once observed, 'Travel makes one modest.' But I think it does so much more. It makes you tolerant, humble and very grateful, too. It makes you recognise that there are many ways of living your life, and no such thing as black and white. By going away on journeys, whether they be short trips or long expeditions, I have learnt to let go of prejudices and accept things the way they are. Away from the constraints of my own society, I have found a contentment like no other, and a reality more authentic than anything I could conceive within the boundaries of those set by myself in an ordinary world.

The more you learn of others, the more you will know yourself.

The stakes are high

While travel gives us an obvious and clearly delineated time to think, it is not enough to adopt this mindset only when travelling. It is a way of thinking and a process that can and should serve us in all areas of our lives. Self-reflection is a must for being honest with oneself, and truly understanding one's own needs and desires.

As humans, our vision of our inner world does not come as naturally to us as our ability to perceive the outside world. If you can learn to understand your inner workings as well as the outer, then you will be emotionally intelligent, have better social skills and be much more motivated. As with anything, some people are born with greater skill at this than others, and one's upbringing also makes a difference, but no matter where you start, everyone has the capacity for self-awareness; like the strength of your muscles, it is something that can be improved upon with practice.

Our self-appraisal mechanism is like a feedback loop: if you welcome information on how you are perceived by others, that is how you will develop. If there is a gulf between your understanding of yourself, your performance and the character you outwardly present, and what others see and know to be true, you will not go as far in life. All of this takes a huge amount of humility and you must have a willingness to put yourself out there, if change is even remotely going to be possible.

Whether consciously or not, I have always enjoyed setting goals for improving myself, but it can be hard and very exposing to put yourself out there and ask for feedback. Alas, it is only by seeking this 'criticism' that you will get to know yourself in the first place, and then have the capacity to grow.

Anyone can do this – it is all about making time for it. Go and write a really honest list of things that you know about yourself. You might have heard of a SWOT analysis. Write down your own Strengths: personal and circumstantial, everything you're good at, as well as your assets. Then list your Weaknesses, and be ruthless here. O stands for Opportunities: what can you see around you that will help you to improve – whether that's a new job, a financial investment, the offer to travel or simply an introduction to a new person. Then finish up with Threats: what is in the way between you and success? Who is the competition for that job you want? Will someone else beat you in the race?

All these things you probably know already, deep down, but sometimes it is important to take the time to reflect on your own situation, where you've come from, and where you're heading. The truth is that if you don't adopt this way of thinking, then you are directly placing yourself as an obstacle between the life you currently have, and the one that you wish you were living. In other words . . .

Know yourself.

That means being truly honest in what we are, and what we want, because without that knowledge we can never know where we want to go next in life. The voyage of self-discovery is often uncomfortable – sometimes downright painful – but what's the alternative? Living a life where you constantly feel lost, and dissatisfied? Personally, I would rather take the path to knowing myself, no matter how painful, because I know what lies at the end of it is a life well lived, and my potential fulfilled. Curiosity, as we'll discover next, is key to exploration.

Remember that to be an explorer means asking questions – of the world, of other people, and most importantly – of yourself.

George Mallory on Everest

2

Be Curious

*The difference between an ordinary life and an extraordinary
one is only a matter of perspective* – Beau Taplin

Escaping the ordinary

Growing up, I was blessed – or cursed – with an insatiable curiosity.
I remember reading *Moby Dick* for the first time as a child, Herman
Melville's words echoing through my soul: 'As for me, I am tormented
with an everlasting itch for things remote. I love to sail forbidden
seas, and land on barbarous coasts.'

I had never sailed a forbidden sea, and I don't remember seeing
anything particularly more barbarous when we left Stoke for trips to
the seaside, but I understood Melville's words as well as I knew my
own name. I believed then – and I believe now – that there was
something in my soul that demanded I explore.

And I mean, *demand*. There was no subtle hint about it. My
head was filled with a raging storm of questions. I couldn't switch
it off, and I knew that the only cure was to find answers. I *had* to
explore.

Of course, I am far from unique in this respect. Think about the
explorers of old.

When George Mallory was asked by a journalist from New York
why he wanted to climb Mount Everest in 1924, the hardy mountaineer responded with perhaps the three most famous words in the
history of exploration: 'Because it's there.'

Travellers and adventurers of all eras have constantly wondered
what is over the next horizon. There seems to be an innate desire in
some people to know and learn more, and an aversion to routine and
stability. I knew at an early age that I was not cut out for a mundane

office job and I wanted to experience something that would provide me with adventure. I always liked surprises.

I also hated the saying 'curiosity kills the cat'. It feels like society is sometimes trying to drum our natural instinct out of us – to make us accept the box that we are born into. If every human had listened to that advice, then there would be no books, no music, no art. I wouldn't have friends all over the world. We wouldn't have medicine. Complacency is what kills the cat – particularly one born with a desire to explore – or as the novelist Paolo Coelho rather astutely put it: 'If you think adventure is dangerous, try routine; it is lethal.'

Danger is not what we should be worried about. It is living a life full of regrets that ought to be a bigger concern. I have thought long and hard about my own reasons for wanting to travel so much. On top of being somewhat in awe of Mr Adam's youthful adventures, I believe a deeper inspiration came from my childhood holidays with my mum and dad.

I was lucky to have parents who were both school teachers. They had long summer holidays and so we were able to travel for at least a couple of weeks twice a year. Because they were teachers, they didn't have a great deal of money, so the holidays usually involved long motorway journeys to places like Scotland, Wales or Cornwall, often with grandparents and extended family in tow – all destined to fill up a caravan, or a great big tent on a muddy campsite. Of course, as a child they were great adventures, which began the moment my mother started packing the suitcases. I knew, even as a toddler, that as soon as we set off, we were into the realms of otherness – an alien world filled with ice creams, audiobooks and service stations.

The anticipation was almost overwhelming. There would be treats and sandcastles and salty water. Coming from the Midlands, seeing the sea was a rare adventure, and it was always exciting to wrestle with my little brother Peter in the back seat of the car for the title of who saw the sea first.

Nothing, though, could beat the thrill of jumping out of the car at our destination, sucking up the fresh coastal breeze, and exploring this new and foreign land. Even at home, my parents fully encouraged me to discover what lay beyond the hedgerows and our garden. They let me build a den in the field and sometimes, if it was warm enough, Peter and I would camp in the garden, imagining we were the very first settlers in this strange environment.

Every time we left the confines of Stoke-on-Trent on a family trip, it felt as if the universe had expanded. A desire to see the rest of the world was cemented in my mind forever. Travel, even as a child, made my little heart sing and kept me wanting more. Those formative holidays, wandering in the Highlands, catching crabs off piers and scrambling around Welsh castles, established a vision of reality beyond my normal comprehension. It was an ethereal existence verging on fantasy, where anything was possible. It didn't so much matter where I went, *as long as I went*, and I immersed myself in the simple experience of being in a new environment away from the routine of home.

Explorers of the South Seas

Thor Heyerdahl was a visionary explorer and anthropologist, who wanted to try and prove that Polynesians originally came from South America and that there was an ancient sea route between that continent and the South Pacific Islands. In 1947, the famous Kon-Tiki expedition was launched. Heyerdahl built a raft made only of materials that would have been available to pre-Columbian society and set sail from the coast of Peru on an epic journey of 5,000 miles across the Pacific Ocean.

Subsequent DNA analysis of the people that inhabit Polynesia has since discredited the Norwegian explorer's hypothesis, and yet his own expedition was a success – proving that with a bit of luck and a steadfast determination, the voyage was indeed possible. What's more, it has been shown that the seafaring people of the South Pacific, whilst perhaps not originating from South America, did in fact travel extraordinary distances on rudimentary rafts. There were no compasses, maps or sextants; they navigated with only the movements and rhythm of the seas and the constellations of the stars.

The Polynesians somehow managed to sail thousands of miles across open water, from the famous Easter Islands to Tahiti, Hawaii and New Zealand, which begs the question: why? As populations on the islands grew, there was a need to search for other islands to inhabit, but the scope of the task was incredible. The islands they colonised seem so tiny when contrasted with the vastness of the ocean that surrounds them, and yet these innate explorers were determined to find out what lay beyond the limited, safe lands that they knew.

The feat of discovering islands in such an expanse of water is nothing short of miraculous, but it was made possible by the insatiable curiosity of those who were willing to set off into the cruel seas against all the odds. Those men and women were rewarded for their inquisitiveness with new lands, rich in resources and devoid of competition, in which to house and feed their families. It is an accomplishment that should leave us all in awe and wondering about our own limits.

Adventure in the genes

Mr Adams used to tell us that only boring people get bored. As a child, I took that as license to let my imagination run free. I was never bored, either in my own company or other people's. I think those early holidays with the family made me aware that there was a whole world waiting to be explored. I knew I wasn't cut out for routine, and travel soon became my passion.

I am far from alone in feeling that way. Travel seems to be a basic human desire, part of our psychological, if not genetic makeup. I get a visceral thrill being somewhere new. For many of us, even the mere anticipation of travel is enough to excite and fill us with a sense of joy, but it begs the question: why are there more than 500,000 people in the air at any one time? Why do we humans, in ever increasing numbers, decide to pack ourselves into flying cylinders and translocate to the other side of the planet simply to take a break?

The vast majority of those half a million flying humans are travelling because they want to, no matter how annoying the airport, or sitting in the middle seat – the excitement and thrill of getting to explore a new place is so addictive. Even as a young man, I wanted to understand where my obsession with movement, newness and adventure came from. What gave me such itchy feet? Why did I appear to have such a deep-rooted and inescapable wanderlust?

The answer may be partly genetic. Studies have shown that a particular allele of the DRD4 gene (which controls our sensitivity to dopamine, effectively the hormone from which we get our kicks), known to geneticists as DRD4 (7R+), is associated with risk-taking behaviours. This allele – dubbed the 'wanderlust gene' – appears to be more prevalent in nomadic communities, who need a drive to explore in order to keep moving and finding new pastures, and less prevalent in settled, sedentary peoples. Globally, around 20 per cent of people carry this allele, so if you feel you are a natural explorer, you might well be a carrier.

That said, the other 80 per cent of us have a drive to explore as well, and the explorer's mindset is relevant to everyone. Regardless of whether it is genetic or not, the desire in many people to explore the world is very real. It is not only me that has a seemingly insatiable need to explore. It is deeply rooted in the human psyche. Ever since early man left Africa, humans have continued to move all around the planet. Cynics might say that they were simply looking for new food and resources, but when there were only a few hundred thousand humans on the planet, there wasn't anywhere near the amount of competition for hunting grounds and grazing that there has been in more recent millennia.

No, there must have been something else that induced these prehistoric explorers to leave the relative safety of their forests and caves to set off into unknown lands and face untold dangers. No doubt the search for food was a factor, as early human populations were at least partially nomadic and their prey came and went with the seasons, but why would a community leave the comfort of the African savannah with its plentiful game and endless opportunities for foraging, to go and seek out the deserts of Arabia, or indeed the harsh, cold wastes of northern Europe and Siberia – let alone brave the perilous sea crossings on bamboo rafts to settle on remote islands in the south Pacific?

One must assume that it was the same motivation that drove generations of successive explorers, conquerors and travellers to venture forth continually in the spirit of exploration to inhabit almost every corner of the earth, from the frozen wastelands of the Arctic to the seemingly impenetrable jungles of South America.

Let's say that there is a wanderlust gene, or at least a basic human need to travel. What could possibly be the benefit to those of us living in the modern world, with all its technological comforts? We live in an age of Google Earth, where the planet reveals its secrets at the spin of a digital globe. We live in the age of TV and online streaming, where we can watch icebergs and steamy rainforests from the luxury of our sofa, so why would anyone bother actually going to such irksome environments themselves?

Why, in England – a country that has long forgotten nomadism – is there still a population of travellers who choose to move from place to place in the same way their ancestors did for hundreds of years? I suppose it might be the same reason that in North Africa and across the Middle East there still exist Bedouin nomads who refuse to settle in one place, in favour of a lifestyle of discomfort and hardship and continual momentum. When there are plenty of occupations that offer financial reward, stability and plenty of other benefits, why do some people choose to become itinerant circus entertainers, roving war photographers or busking yoga instructors? Out of curiosity, that's what.

I could read a thousand studies, or interview a hundred scientists, but the truth of the matter is this: I *feel* I have to explore, and therefore I must. I know myself, and therefore I can match my external action with my internal feelings and thoughts. I *have* to be curious, and explore, because it is who I am.

Sound familiar?

A bit of distance

Solvitur ambulando – 'Everything is solved by walking' – is a phrase first attributed to the ancients, who used to say that all temporal and spiritual conundrums could be figured out by taking a long walk. It was an idea that has been passed down through the ages – from the early Christian pilgrims to modern philosophers and travellers. It is the theory that removing yourself from a problem for a while will give you the headspace required to work things out.

When I was a soldier, one of the key skills we were taught for command was how to detach ourselves from what was happening around us; with all the bangs and the shouting, it was easy to get sucked into what was right in front of you. Through being put into challenging situations, we developed the skill to take a mental step back from the chaos and the noise in order to see the bigger picture, and work out a plan to overcome the situation in front of us.

Travel can be the 'step back' that gives us that detachment. When a problem seems to be 'in our face', it can be difficult to think of any solution to it other than fight or flight. By contrast, when we escape from the place where we spend most of our time, our mind is suddenly made aware of all the errant ideas we had suppressed. We start thinking about obscure possibilities that never would have occurred to us if we had stayed back home. Furthermore, this more relaxed sort of cognition comes with practical advantages, especially when trying to solve complex tasks.

Take the 'Duncker problem', a cognitive test published in 1945, when people are given a cardboard box containing a book of matches, a few drawing pins and a wax candle. They are told to figure out how to attach the candle to a piece of cork board on a wall, so that it can burn properly and no wax drips on to the floor. Nearly 90 per cent of people pursue the same two strategies, even though neither strategy can succeed. They try to pin the candle directly to the board, but this causes the candle wax to shatter; or they try to melt the candle with the matches so that it sticks to the board, but the wax doesn't hold and the candle falls to the floor.

At this point most people surrender, assuming that the puzzle is impossible and the experiment a waste of time. Only a slim minority come up with the solution, attaching the candle to the cardboard box with wax and then pinning the cardboard box to the corkboard. Unless people have an insight about the box – that it can do more than hold drawing pins – they will waste candle after candle. They'll repeat their failures while waiting for a breakthrough. This is known as the bias of 'functional fixedness', since we are typically terrible at coming up with new functions for old things. That is why we are so surprised when we see an everyday object used in a way for which it wasn't designed.

Researchers reported that those who had lived abroad were 20 per cent more likely to solve this problem than those who had never lived outside their country of birth. Why? The experience of another culture endows us with a valuable open-mindedness, making it easier to realise that a single thing can have multiple meanings. Consider the act of leaving food on the plate: in China this is often seen as a compliment, a signal that the host has provided enough to eat. But in America the same act is a subtle insult, an indication that the food wasn't good enough to finish.

In another study, an American psychologist found that people were much better at solving a series of insight puzzles when told that they came all the way from California and not from down the hall in Indiana. The subjects considered a far wider range of alternatives, which made them more likely to solve the challenging brain-teasers. There is something intellectually liberating about distance. In reality, most of our problems are local – people in Indiana are worried about Indiana, not China or California.

This leaves two options: find a clever way to trick ourselves into believing that our nearby dilemma is actually distant, or else go somewhere far away and then rethink our troubles when we get back home. Given the limits of self-deception – we can't even tickle ourselves properly – travel seems like the more practical possibility.

These cultural contrasts mean that open-minded travellers are receptive to ambiguity, more willing to accept that there are different (and equally valid) ways of interpreting the world. This in turn enables them to expand the circumference of their cognitive inputs, as they refuse to settle for their first answers and initial guesses. Of course, this mental flexibility doesn't come from mere distance; increased creativity appears to be a side-effect of difference – we need to change cultures to experience the perplexing assortment of human traditions.

The same details that make foreign travel so confusing – Do I tip the taxi driver? Where is this bus taking me? – turn out to have a lasting impact, making us more imaginative because we're less blink-ered. We are reminded of all that we don't know, which is a lot; we are astounded by the constant stream of surprises. Even in this globalised age, slouching toward homogeneity, we can still marvel at all the earthly things that aren't included in the guidebooks and that certainly don't exist back home.

Consider too the impact that distance has on our emotions. How do you feel when you watch a movie on a flight or in the cinema, free from the subconscious restraints and pressures of your home? Why do so many people find love on one-week getaways, when they tell themselves that they can never find it at home? Distance can free us from the self-imposed cages that we have built at home through our routines, the norms of our society and the expectation and opinions of others.

Don't ask, don't discover

Albert Einstein once said: 'I have no special talents. I am only passionately curious.'

We are curious animals, and we forget this simple fact at our peril; it's the key to self-knowledge. We are far more predisposed to ask questions about the world around us than we are to question ourselves, so if we don't do the former, what hope do we have of doing the latter? Exploring gives us the opportunity both to see ourselves through a different lens – comparing our own upbringing to that of someone with a very different lot in life – and to learn things about ourselves by facing challenges we would otherwise not encounter.

This could be anything from climbing a mountain to using a hole-in-the-floor toilet – the important thing is that it is new, and outside of our comfort zone, because no matter how big or small the experience may seem, it contributes to our personal growth. Simply realising that there is no right way or wrong way to live is in itself hugely powerful. Without that knowledge, we succumb to tunnel vision and group think, which shrink our horizons dramatically.

Social/cultural anthropologists, who study different human cultures, have a phrase for this kind of perspective: 'cultural reflexivity'. It describes the ability to recognise that the way we perceive the world is only one of an infinite number of world-views, and that, for example, Western materialism doesn't seem any more strange to a Tibetan monk than a belief in reincarnation might to a Wall Street banker; but it *is* just as strange. By experiencing other ways of viewing the world, we can understand that our own world-view (and its associated stresses and challenges) is in many senses arbitrary.

Without reflexivity, all we can really do when we visit another place or culture is to compare that way of life with our own. This is flawed, because each way of life is unique and rests on its own particular values and life prospects, and we lose any chance we had of learning something from other people. Rather than compare, we need to learn to place ourselves fully in someone else's shoes when we travel. We do this by asking questions before we make any assumptions, and therefore we open ourselves up to infinite veins of wisdom and allow ourselves to tap into new perspectives. How many of our problems could be solved by finding a new way to look at them?

It is not enough simply to get on a plane or visit some unexplored part of the world: if we want to experience the creative benefits of travel and exploration, then we have to rethink its *raison d'être*. Most people escape to Paris *not* to think too hard about anything much; but rather than gorge on that buttery croissant, consider a thought for the person who baked it, and how they might view the world differently to a person from across the channel. As a Brit, we can find the idea of shorter French working hours 'frustrating and lazy', but place yourself in their shoes. What do those extra hours away from work bring? More time with the family, or time to read – the ability to separate work from life?

There is a reason behind everything, and to imagine that we know it without asking questions is foolhardy at best, and jingoistic at worst. Put aside ego – both personal and national – and ask *why*. Then ask again, because the first answer is rarely the correct one.

The truth is that – as nice as it is – you don't need to go to Paris or Peru to do this kind of exploration of the mind and our assumptions. There are many ways we can expand our horizons without needing to leave our home town, or even our home. It could be as simple as trying new cuisines; perhaps visiting community centres, or places of worship that you don't normally associate with, and asking the people there about their beliefs and cultures; or reading books based on the beliefs of other cultures (even other religion's holy texts, if done respectfully) and ethnographies.

Even watching a classic Bollywood film – given how different the traditions of filmmaking are in India from the West – can have a transporting effect, and provide distance from your immediate life. Something as mundane as changing your route into work occasionally will show you people and parts of your town you've not seen before. Be honest with yourself – how well do you *actually* know your town, or your neighbours? For that matter, how well do you know your family, or have you fixed assumptions based on their role in your life?

The larger lesson here is that our thoughts are shackled by the familiar. The brain is a neural tangle of near-infinite possibility, which means that it spends a lot of time and energy choosing what *not* to notice. As a result, creativity is traded away for efficiency; we think in literal prose, not symbolist poetry. A bit of 'distance' (geographical or mental), however, helps loosen the chains of cognition, making it easier to see something new in the old, and the mundane is grasped from a slightly more abstract perspective.

David Livingstone

On the outside wall of the Royal Geographical Society in London is a statue of perhaps the most iconic of all British explorers. His name inspired one of the most famous lines in the annals of Victorian exploration, uttered by another great adventurer, Henry Morten Stanley, who was sent out to find the lost hero: 'Dr Livingstone, I presume?'

Dr David Livingstone was of humble Scottish origins, growing up in Blantyre in a workers' cotton mill to impoverished parents. Against all the odds, he worked hard to become educated and became a medical missionary. He was a man of great conviction and determination to improve not only himself, but all those around him. As a missionary, he was sent to Africa and was a vociferous campaigner against the slave trade. Unlike many explorers of his age, he showed a desire to understand the people he met there, rather than simply convert or exploit them. He learned the local languages and immersed himself in African culture and customs, all the while never losing his zest to explore this uncharted land.

He undertook a number of great expeditions, travelling along the Zambezi river into Central Africa, further than any white man at the time. He also went in search of the source of the Nile and in doing so mapped out much of the Great Lakes region of Tanzania, adding a wealth of new information to the geographical understanding of the time.

That said, he never succeeded in many of his tasks. The source of the Nile remained elusive and his missionary efforts were ultimately futile. But despite the fact that his achievements were far less glorious or impressive than many of his contemporaries, it is his name that stands out, thanks in my view to his endless curiosity about the world. He followed his heart to the extreme and regardless of the outcome, he did what he felt was right.

He was a stern believer in personal change and choice, and always encouraged others to choose the right path, never forcing anyone to his own ideas. Unlike some later Victorian explorers who believed in racial superiority, Livingstone called that attitude 'the most pitiable puerility' and was a staunch believer in equality and human potential. He was a patient, kind and intellectually curious explorer, who embodied the virtues to which most can only aspire.

Seeing things in a new light

When we travel, there is usually a goal behind it, even if it is only to rest and do nothing. However, we often find something that we had no idea we were looking for. This makes sense when we view travel as a learning opportunity, dictated by our circumstances back home. We might be stressed, so go away looking for peace and quiet; we might be interested in history, so go somewhere in search of ancient monuments and landmarks.

It is important to have a goal like this in mind – after all, it is part of being curious and the point of travel in the first place – but on top of these goals, we need to be ready for experiences that we were not expecting. This is how travel and exploration can be truly transformative; they introduce our minds to things and events that previously they couldn't even contemplate.

I've always loved elephants. Ever since I was a kid I wanted to see them in the wild. I was inspired by David Attenborough's documentaries and loved the fact that, against all odds, these enormous gentle giants still roamed across the wilds of Africa. I wanted to see them with my own eyes, So when I was eighteen, I set off on a very clichéd backpacking trip around the world. It was 2001 and going off on a gap year had suddenly become all the rage. I had never travelled on my own before, and it was terrifying.

First stop was South Africa. I got a flight to Cape Town and took a bus along the coast via Durban all the way to the Kruger National Park. I was very much out of my comfort zone, but I put on a brave face, pretending to know what I was doing. At first, I kept myself to myself. There were lots of backpackers in their mid-twenties and I felt like a child, tagging along on a grown-ups' holiday. Then I realised that everyone else was as nervous as I was.

I experimented by talking to strangers and forcing myself to interact. I learned card games and listened to stories and observed how other people acted. I slowly grew in confidence and began to hitchhike as a way of meeting random people. In doing so, I was invited into strangers' homes and shown incredible hospitality and kindness. It established a firm belief in human nature, which I have taken with me ever since.

I found that by putting myself in new situations I would be afforded new opportunities, and that was how I came to be invited on a free safari, where I got to see elephants in the wild. But for me, the biggest reward wasn't seeing the elephants, although they were awe inspiring; it was understanding more about other people, learning new ways of doing things, and discovering more about myself and my own potential.

In hindsight, it was a pretty simple and self-obvious truth that I would learn more about the world in general, and myself in particular, from meeting strangers along the way than I would from seeing a bunch of elephants, but I had to take the trip to grasp that. In itself, that's a great example of how blinkered our brains can become if left to the same old, familiar routines. We forget that we share our planet with seven billion other people, and their lives are so hugely different from our own that each of them has something new and intriguing they can teach us. The trick to netting as much of this collective wisdom as possible is exploring at home and abroad with an open, inquisitive mindset.

Sir David Attenborough

David Attenborough was born in Isleworth in May 1926, and grew up in Leicestershire. From an early age he had a voracious appetite to learn about the world around him, and was fascinated by nature, wildlife and fossils.

Attenborough's curiosity led him to the field of nature filmmaking, beginning with the three-part series The Pattern of Animals in 1953. At time of writing, almost seventy years later, he has yet to hang up his camera, nor has his curiosity been dulled. In 1957, Attenborough founded the BBC's Travel and Exploration Unit, from which he commissioned a huge range of programmes that he could share with his fellow man.

Through the course of his career, Attenborough essentially created the genre of the nature documentary as we know it, and he has become virtually synonymous with this type of programme. He has also become known as a champion of environmental causes and is an ardent campaigner. But Attenborough's curiosity and interests stretch far wider than the field of nature and conservation; during his time as a controller of the newly created BBC Two during the 1960s, he committed to broadening the range of content beyond that shown by traditional broadcasters, commissioning programmes as diverse as Match of the Day and Monty Python's Flying Circus.

Sir David Attenborough has been given more honorary degrees by British universities than any other individual. He also has an extensive list of titles, having been made CBE in 1974, knighted in 1985 and made a Fellow of the Society of Antiquaries in 2007. In a dazzling career that is testament to the rewards of curiosity, perhaps one childhood incident stands out above the rest: as an eleven-year-old, the young David responded to a request from University College, Leicester, for newts for the zoology department, which he offered to sell to them for three pence each. His source of the newts, unbeknownst to the university, was a pond five metres away from the faculty. A bit of curiosity about the world around you pays dividends!

Look beyond the horizon

It is amazing what bits of information we hear in our childhood, but don't fully come to appreciate until we're older. My grandmother used to say that a change is as good as a rest, which was hard to understand as a child. Now, I couldn't agree with her more.

I've already mentioned a few of the reasons that we explore. The truth, of course, is that there are as many reasons to travel as there are people on the planet, and on top of that you can layer the overlapping circumstances that arise as we do our best to navigate our lives. And yet, I believe there are 'themes', if you will, that must unite us in our exploration, not least of which is the desire for a sense of freedom and liberty, unique to the process of leaving behind one's norm and setting off to somewhere different.

Exploring new horizons is a mindset, as is applying curiosity to all that you do. It's about pushing your personal boundaries and taking the time to see and do new things, wherever in the world you may be. My own childhood emotions of excitement, adventure, a sense of freedom and joy are what still compel me to travel, as I'm sure is the case for many people who pack a bag, or simply walk out of their back door and into their home town that still holds surprises.

There is a sensation that comes with the unfamiliarity of places. A visceral feeling as the senses are crammed with new smells and flavours. You don't just see a place when you travel – you smell it, you touch it and feel it. The hot sand beneath your toes or the wet jungle air in your lungs. Travel is immersive. We go to these places expecting that they exist for our admiration, but quite often it is we who are consumed by the experience. The mosquito that drinks our blood. The villager who makes a sale and sits with the grin of a man who has charged you four times his normal price.

We are not set aside from each new place that we visit. We are a part of it, and we will remain so even after we leave. Our interactions will be remembered by the people who call it home. They were introduced to the new as much as you were.

I love the human connection when I travel. I also like surprises, and the feeling of anonymity of being in a different country. It makes me feel invisible and free. I especially like those sensations when I'm alone. Usually the first thing I do when I get to a new city is go out and walk, with no particular destination in mind; I will happily get lost. I walk the streets without a map and the simple process of walk-ing aimlessly gives me time to reflect, check in and get to know myself once more.

I go to far-flung places to do this, because I grew up travelling the world; I have set my boundaries quite far to start with, so to push them I need to go a long way away. If most of your life has been spent in one country, even one town, then good news: you've got less far to go, at far less cost and effort, to stretch your own boundaries and gain a sense of the unfamiliar. Take it one step at a time, and see where it takes you.

And while not everyone can or wants to 'travel', anyone and everyone can 'explore'. In his book *A Journey Around My Room*, Xavier de Maistre describes the experience of 'travelling' to his sofa and seeing it with fresh eyes, studying it deeply and celebrating the sensation of exploration and discovery without needing to leave his own house.

The same benefits that a traveller gains from exploring the world around them – perspective, novelty, challenge – can be gained simply by adopting a curious, exploratory attitude towards the familiar, everyday things around us. With the explorer's mindset, we can see ourselves through someone else's eyes, and come to know our own limits and aspirations better as a result. In 2020, with prolonged lockdowns, how many people have fully come to know their home for the first time, and the neighborhood that surrounds it? We don't have to wait for a pandemic to make us curious about the place we call our own.

Curiosity opens new opportunities and perspectives to us. A curious mind is the most powerful thing that you own. It broadens our horizons and enables us to reassess our lives in a new light, providing distance from our problems. But more than that, it sets us on a journey towards unlocking our fear of the unknown. Overcoming our fear, in all contexts, is the subject of the next chapter.

Amelia Earhart

3

Do What Scares You

The cave you fear to enter holds the treasure
you seek – Joseph Campbell

Take the leap

One hot morning in 2001, I set off from my hostel in Northern Queensland in search of an elusive waterfall. It has grown in popularity in recent years, but back then it was barely on any maps, and only the local people knew the way. The waterfall had attained almost mythical status among young backpackers, because it involved several hours hiking to get there, and there were no organised treks. I'd never seen any pictures, but the two or three people I'd met who had made the journey had described it as one of the most beautiful they'd ever encountered – consider me sold!

I packed a small backpack including drinking water, sunscreen, a few snacks, an old film camera and a compass that I always carried. The hostel owner had given me directions on how to get there through the forest, but no one else seemed interested in going, so I set off alone.

At least that was the plan.

I had barely started walking down the road that led into the forest when I noticed a little dog following on behind. I stopped to pat it and saw that it was a beautiful grey Staffordshire Bull Terrier and – being from Staffordshire myself – I took it as a good omen. I tried to shoo him away, because I didn't want the poor thing to get lost, but as I carried on walking down the path, the Staffie seemed determined to follow me.

The forest was hot, and seemingly deserted except for myself and my new companion. I could hear the ocean on the other side of the trees, but no other people. Then, as the path petered out into nothing, I realised that I was lost. I decided that the best route would be to walk towards the sea and follow the beach north. After hacking my way through the bush, I reached the deserted shoreline, where palm trees cast a short shadow over the brilliant white sands. It was a scene of such beauty that I could hardly believe it existed; the kind of place reserved for postcards and computer screensavers.

Spewing out of the jungle to my left was the estuary of a small river. It seemed to cut a swathe through the sand like a green snake, before pouring into the shimmering ocean. The estuary was perhaps six metres wide, too wide to jump across, and there was no way to tell how deep the water was, or what lurked within – this was Australia, after all.

I stared at the water for a while, but my intense concentration did not conjure up a bridge. So I thought about looking to see if it got any more narrow inland, but that was a no go as I knew that to cut through the thick forest would be virtually impossible.

It was then that I was delivered a sign.

A rusty sign, mind you. It had been hammered into the sand. The sea salt had faded the yellow paint, but its message was clear to see – there was a black drawing of a crocodile. It looked almost comedic, with a curly tail, the kind you see in children's books, not at all menacing. And yet I knew there was nothing laughable about the situation. I'd heard all about the infamous saltwater crocodiles and their fearsome reputation. Some were so massive, they would eat sharks that dared to swim up into the river mouth. What was I going to do?

As I stood on the hot sand, a few stories floated up to me from my past. I remembered the story that Mr Adams had told us at primary school. I couldn't recall the moral of it, but by taking a risk he had found his way to the other side of the railway tracks and made it home. And then I remembered the cliff, my dad and how he'd given me the confidence to jump off it alone.

I looked at my watch. Forty-five minutes had passed while I'd been daydreaming in indecision. The dog was getting impatient now, looking at me as if I were a coward. I decided that there was a simple way to commit myself to action, and threw my rucksack across to the other side of the waters.

Shit, I thought to myself, *there's no going back now*. The Staffie knew that. Wagging his tail, he jumped straight into the river, swam to the other side and gave his coat a shake all over my bag.

Shit.

I walked back a few yards, determined to run up and jump, so that I would spend the least amount of time possible in the water. I came forward, but then stopped at the water's edge: I was scared. I didn't want to die, and particularly not in the jaws of a prehistoric monster. No one would even find my body, only my bag on the sand and a stoic Staffie, who would probably not make the greatest witness at my inquest.

I took a deep breath, and made myself think about the odds. There are a lot of crocs in Australia, and a lot of people, but only a few get eaten a year. The odds were on my side, right? I took a few steps backwards then ran, and launched myself into the air, bracing myself for the hungry jaws that awaited me . . .

But there was only a splash, and a thrashing young man, and then I was out on the other side – and alive. I felt exhilarated and foolish all at once, and then I realised that I'd have to do it all again on my way back. Oh well, I'd cross that bridge – or lack of it – when the time came.

I kept walking up the beach with my companion, and after another hour we reached the waterfall. It was every bit as beautiful and unspoilt as it had been described to me, and the fear I had felt in getting there was washed away.

DISCLAIMER: Now I'm not advocating recklessness here, and I'm sure that there are plenty of parents who will be tutting with desperation at the thought of their gap-year, adventuring offspring taking silly, unnecessary risks because I told them to hop over croc-odile-infested waters.

But . . .

Some measured risk is a crucial part of our development as we grow up. It is something to be sought out, rather than avoided.

Mawe Manhood

There are many different manhood rituals around the world, but one of the toughest and most painful has to be performed by the Mawe. The tribe live as hunter gatherers deep in the Amazonian jungle of Brazil.

It is an unforgiving environment, and they believe that for the younger boys to become men, they must prove that they can endure the most painful experience there is: the sting of the Paraponera clavata, also known as the bullet ant.

It may not surprise you to hear that the bullet ant takes its name from the fact that its sting produces a feeling like being shot. The sting is 4 out of 4 on the Schmidt pain index, with only the bite of a tarantula hawk wasp given the same score in the insect world. It is said to be excruciating and wildly painful, with waves of burning pain lasting up to twenty-four hours. After covering them with a natural sedative, the Mawe weave hundreds of these ants into a pair of gloves made from leaves and vines. Naturally, the ants' stingers are left to face inwards in these mittens of misery.

For the manhood ritual the young boys must place their hands into the ant glove for five minutes, receiving hundreds of stings. Through the process the boys experience muscle paralysis, disorientation, violent shaking and even hallucinations. But once it's over, it's over, right?

Actually, no. To be accepted as men, the boys must perform the ritual up to twenty times, which may take place over months, or years. Once the men have faced the most excruciating pain the jungle has to offer, they have faced their fears and are able to hunt courageously in the rainforest as a man.

Embrace risk

I am not one for putting my hands into gloves of bullet ants, but a big part of my own mindset *is* about allowing curiosity to get the better of fear. There are amazing things, places and people around us in the world, but so often a nagging fear of doing what others find odd, of social rejection, or maybe even financial ruin and physical danger, prevents us from discovering something new. Through practice and confidence, we can all get a grip on our fears and approach the unknown with curiosity and interest.

Believe it or not, I have an innate fear of heights. I'm terrified of flying, in fact, even though I know rationally that it's safe. Because of this, over the years I've forced myself to face that fear. I joined the Parachute Regiment, I've climbed mountains, and I've tried to learn paragliding. I'm still scared of heights, but each time I push myself, I become a little less so. Explorers are sometimes described by journalists as fearless, but it's not true. I think we misappropriate the word, because really those who don't fear are probably either mentally deficient or psychologically maladapted. In fact, there is evidence to suggest that psychopaths have a reduced startle response and that their fear receptors are dulled.

Fear has been given a bad rep. It isn't something to be avoided; rather, it is our life's mission. It's our duty to overcome our fears, whatever they may be. I believe that failure to do so is an abject failure in life itself. With that being the case, what do we have to lose?

Contrary to what we might read in the news, we live in some of the safest times in all of human history. People live longer, we have amazing technology and medicine, and fewer people die from illness and war than ever before. And yet we seem to worry more, and be scared of everything.

I try to remind myself every day that the world is not as dangerous as the media plays it up to be. The threat and fear that is transmitted to our screens twenty-four hours a day is not a fair representation of the seven billion people living in amazing places across the planet. I've found even in the most war-torn countries that fighting and danger exists for only a small amount of people, for a fraction of time. Equally in the most hostile places, you're likely to find the most incredible hospitality. It is up to us to accept our own powers and make up our own minds, and not leave it to someone else.

Fear is a natural human instinct that has evolved as a tool for survival. The fight or flight mechanism in our brains that has been a part of our reptilian core for millions of years still dictates how we deal with stressful and potentially dangerous situations. No matter how clever we think we are, we're still evolved apes, and fear is an essential part of our psychological makeup.

Fear is contagious. This is how terrorism works. It kills very few, but terrifies billions, forcing nations to spend their security budget on preventing panic. Fear can collect and pool in an area, becoming part of the collective consciousness, as it has recently in Paris and

London. This is not just in our minds – there is evidence suggesting that humans emit alarm pheromones when under stress, which send out olfactory signals that can be detected by other humans.

Greed may be the root of all evil, but I believe that fear lies at the root of all greed. One of the seven deadly sins, it is by definition insatiable, making it all the more lethal. Greed amounts to taking more than your fair share, or grasping and clinging onto something that you are scared of losing or not having enough of. Quite frankly, as evolved apes, we cannot always resist these animalistic urges, which explains why most of us have been greedy at some point in our lives. Taking more than one's portion is the subconscious ape telling us that we need to have this or that to survive: more food, more sex, more money, more power. We are basically scared of not having enough – so we take more than we need. We become desirous, and lack control over ourselves.

Having evolved from animals that would be only sporadically successful in bringing down prey, dogs have an instinct to keep eating until their bellies are completely full. In an age where human owners provide all their food for them, domestic dogs are unable to suppress this instinct – meaning that if their owners succumb to the puppy-dog eyes and overfeed them, they can easily become overweight out of a primal fear that their next meal may be some way off.

Thanks to similar instincts, humans often overeat, and also overindulge in lots of other areas. Our mind loses itself and acts irrationally, defending its actions and saying things such as: 'I need to have that sportscar to feel fulfilled'; 'I need to invade that country to feel safe'; 'I need to kill that person before he kills me'; 'I need to eat that animal, otherwise I won't have enough protein.' Greed is the manifestation of the ego, which is fuelled by fear. In some ways, fear is a manifestation of our greed for safety, which we often think we need more than is good for us.

Greed is the concept of doing more harm than good, or taking more than you give. From an ecological perspective, this couldn't be clearer to me – our greed looks set to be our downfall. It is an exaggeration of fear: the belief, subconsciously or not, that if you don't accumulate, you will not survive. We have become obsessed with material things and with convenience, but it is melting our glaciers and destroying our planet. It is something I have seen first-hand, out in the Himalayas and the Amazon for example, and it's something in which we are all complicit. I don't know what the answers are, but I've become conscious of this greed in my adult life – just because you can do something, doesn't mean that you should.

When we act selfishly or greedily, we make decisions that in the long run are bad for ourselves and bad for society. I'm a firm believer that deep down we all know what is right or wrong, and what we want, but we get a bit lost on our journey to find it. Everyone makes mistakes, and everyone has the opportunity to learn from them. The problem is that when we let our ego control our emotions, we end up holding grudges, fighting with our friends, starting wars or destroying the planet.

Alexandra David-Néel

Alexandra David-Néel lived the life that many explorers dream of. She was an independent, highly intelligent and seemingly fearless woman, who defied all expectations of her time.

Born in 1868 to Belgian-French parents, she ran away from home aged fifteen to learn Buddhism and the ancient arts of the East. An accomplished linguist, she first had to study English so that she could read the translated texts in the British Museum. Determined not to settle for a normal life, she lived with her lover out of wedlock and announced that children would not suit her wandering lifestyle.

In 1924, disguised as a poor pilgrim, she managed to sneak into the forbidden City of Lhasa in Tibet, which was then out of bounds to foreigners on pain of death. She ended up spending two months inside the holy city, where she studied philosophy and became a master Buddhist herself.

Over the course of her life she wrote thirty books about her adventures, as well as academic tomes on history and religion. She travelled her entire life, speaking a dozen languages fluently. At the age of sixty-nine, she travelled through China, where she witnessed the brutal Sino-Japanese war and saw first-hand the horrors it entailed. Her curiosity, though, was never sated, and she was a brave advocate of seeing the world with one's own eyes, rather than simply being a passenger in life.

She once said, 'Who knows the flower best? The one who reads about it in a book or the one who finds it wild on the mountainside?'

Have the courage to live the life of your own making.

Dare to fail

I was only eighteen when I was in Australia, leaping into rivers, and it will come as no surprise to you that teenagers are more likely to take risks than adults or younger children. A teen brain is still undergoing maturation, which makes a lot of us more likely to seek new and novel experiences. The brain has a bias at this age, weighing positive experiences more heavily and negative ones less so. My adult self might have acted differently with the crocodile conundrum.

Effectively, as a teenager you are more accepting of a consequence that is as yet unknown to you. When you capture this mindset and tap into it, you can further broaden your horizons in adulthood. Things are rarely as dangerous as they seem at first glance, and risk is often rewarded. We can always make excuses about why we didn't want to do something, but with that attitude, what would we ever get done, or explore? We would simply stay put in our safe little bubbles, never growing. What a waste.

There is no reward without risk, and certain considered risks ought to be embraced in order to reap rewards. This is all part of being an explorer.

Fortune favours the brave.

Yes, the risk can be scary – it can be downright *terrifying* – but you have to get used to this uncertainty if you are ever to change, innovate or grow. When you take a risk, your brain has an increase of dopamine, and if you learn something new as a result, the neurons form stronger connections. Boldness pays dividends. By taking chances, one can achieve things that other people can't, because they didn't have the bravery to do whatever was required.

On the Victoria Embankment of the River Thames in London, outside the Ministry of Defence, there is a stone plinth topped with a bronze statue of a fearsome-looking beast called a 'Chinthe'.

Resembling a lion, but with the features of a dragon and a dog, this mythical beast is usually found as a stone effigy guarding Burmese temples and pagodas. But during the Second World War it became the symbol of British and Indian Army Special Forces units, who saw action against the Japanese in the jungles of Myanmar.

Nicknamed the 'Chindits' and led by a maverick brigadier called Orde Wingate, these guerilla fighters specialised in attacking the enemy deep behind the front lines. Wingate, a rebel by nature, was very much a budding adventurer himself. As a young officer, he once took six-months' leave to cycle across Europe and Africa to reach his new posting in the Sudan, where he established himself as an expert in desert warfare. He mounted numerous successful operations to capture slave traders and ivory poachers, and in his spare time surveyed North African archeological ruins and published his work at the Royal Geographical Society.

It was the knowledge gained during these early expeditions, and his innate curiosity and ability to overcome fear, that gave him the confidence to become an effective Special Forces commander in Myanmar. The Chindits operations took inspiration from those early forays in Africa and included long marches through extremely difficult terrain, blowing up enemy lines of communication, expedition-style surveying and using guerilla tactics to fight larger forces.

Wingate and his band of irregular fighters were famed across East Asia and feared by the Japanese invaders. It was precisely because they took lots of risk (which was often criticised by other military units) that they had such an effect in undermining the enemy morale and boosting their own forces' efforts.

On the statue in London is written the motto of the Chindits, adopted from the words of Admiral Horatio Nelson: 'The Boldest Methods are the Safest.'

By encouraging risks, we also have a duty to accept mistakes. We must foster an environment where they are not simply tolerated, but encouraged. I like to call this way of thinking 'fail fast, fail early'. This means creating fertile ground for experimentation; we owe it to ourselves to try things out and accept that they might go wrong. There is no time like the present to have a go and get the learning out of the way.

So many people shy away from this foggy arena of mistakes and failure, but only by pushing past your boundaries and limitations can you ever increase them. You are unlikely to achieve greatness if you're safely at home with the doors closed, doing the same thing every day. Routine and regularity are the enemy of pushing boundaries and taking risks. But those who *do* go there – into the arena of risk – can exploit the valuable lessons that come with learning from their mistakes.

Take a look around you in your own life, and at the world at large. I bet that the people or organisations who have grown the most in a short space of time took risks, and those that have stagnated did not. Now, of these, who talks about their stumbles as well as their successful leaps? I believe that organisations and individuals can be divided into two categories: those who conceal their errors and those who confront them.

The aviation industry is a leader in this arena of learning from mistakes. It learns from and interrogates errors, rather than being threatened by them. They install small, robust black boxes in all passenger planes, which record huge amounts of data about the flight and are designed to withstand crashes. Famously, they are rescued from the rubble if an aeroplane has an accident, but they are also used more regularly than this – for instance, if there has been a near miss between two planes, or one has come close to running out of fuel while in a holding pattern, the aviation companies will examine the information. The recorded log, both from the point of view of machinery and people, gets analysed in depth.

The industry has developed a culture of making improvements and reforms, ironing out weaknesses in the system rather than blaming individuals for errors. In aviation, mistakes and accidents are not concealed, but are seen as valuable learning opportunities – crucial in a sector where mistakes come at the cost of human life.

If we fear failure, or allow our fear to stop us taking risks, we cut ourselves off from every chance we have of learning and growing. Organisations and entire industries that are ruled by a fear of failure will stagnate and, ultimately, die out. Leaders who do not create and foster environments in which experimentation is encouraged, where failure is viewed as a learning opportunity, will sacrifice the innovative capacity of their teams and be overtaken by their competitors. Individuals who let fear govern their actions and avoid risky activities over time will not only fail to push their own boundaries, but shrink in on themselves, as the risk-taking, experimental instincts of youth disappear, in favour of the safe predictability of the familiar.

Amelia Earhart

Like many explorers, aviator Amelia Earhart clearly had the desire to win a place in the annals of history. She also had an extremely healthy appetite for risk. As a child, she fashioned a makeshift roller coaster out of crates in her garden, hungry for thrills even at a young age. She loved going to stunt flying exhibitions, watching the hair-raising loop the loops with delight. During one such event, a pilot saw her watching the show and intending to give her a bit of a fright, he dived his aircraft down towards her, swerving only at the last moment, but brave young Amelia sat tight. She said later that she believed the little airplane had said something to her as it whizzed by, enticing her into a life of aeronautical adventure.

In her early twenties, she finally went up in a plane herself. She reflected that she knew immediately – a few hundred feet off the ground – that she had to learn to fly herself. Amelia did not observe conventional limits, which is all the more remarkable given that she was born in Kansas, USA, before the turn of the last century. A time when a great number of limits were a feature of a woman's life.

Flying had not even been around for twenty years, but that did not deter Earhart, who dropped out of nursing school and tracked down one of the best aviators of the time. Neta Snook was the first woman to have a flying school and she agreed to teach Amelia – for a fee, of course, and after a lot of begging. The airfield was a fair distance from Earhart's home, a long bus ride followed by a six-kilometre walk, but she stuck to her training and it paid off; after only two years, she became the sixteenth woman in the US to be issued with a pilot's license. Earhart worked multiple jobs to support her dream – as a lorry driver, a photographer and a stenographer – saving every cent she could spare. After six months she was able to afford a biplane of her own.

From then on, Amelia Earhart began setting world records, and soon became a celebrity. By 1932 she became the first woman – and the second person – ever to fly across the Atlantic. It was a dangerous and testing flight that included an emergency landing. A few years later, Earhart set her sights on the ultimate goal; she wanted to become the first woman to circumnavigate the globe by aircraft.

It was an enormous risk with a huge amount of unknowns: she would be vulnerable to icy conditions, heavy winds or technical problems with the plane. She would need to fly 47,000 kilometres around the equator, which could take weeks, but Earhart was well aware of the hazards. Accompanied by her navigator, she took off from Miami in 1937, and made good headway on the first legs of her journey. A month later, Amelia Earhart had made it to Papua New Guinea, where she took off for the next section of the journey – aiming for Howland Island in the Pacific.

But the formidable explorer never made it. A few distress signals were detected, but the plane and Amelia were never seen or heard of again. In a testament to her popularity, the President of the USA, Franklin D. Roosevelt, sent out a search party. It went on for two years until the teams were forced to give up, and she was declared lost at sea.

Though it is not known what happened to her in the end, Amelia was a trailblazer who has left a legacy of pushing the boundaries and being bold. I read her inspiring tales as a young lad and I have tried to adopt her attitude towards taking risks. Amelia herself said, 'Flying may not be all plain sailing, but the fun of it is worth the price.' She saw risk as something that was an important and necessary part of life, and something that would reap an immense reward – the freedom and euphoria of flying your own aircraft.

As she put it: 'Decide whether or not the goal is worth the risks involved. If it is, stop worrying.'

Keep pushing boundaries

In 2004, at the age of twenty-two, I hitchhiked along the ancient silk road from Europe to India. It was one of the most terrifying and yet exhilarating times of my life. I was fresh out of university and full of the zest for life. Unfortunately, my bank account wasn't full of anything at all. In fact I was attempting to travel for almost five months on just £500, which as you can imagine was stretching things a bit thin.

I decided that I had no choice but to believe in the inherent goodness of people. I was travelling overland all the way from England to the Himalayas and had already crossed Europe, Russia, the Caucasus, Turkey and Iran. By the time I reached the Afghan city of Herat I was out of cash, and had no hope of more till I reached Kabul. I hitched rides in beat-up minibuses and sometimes walked, and one day I reached the small town of Chaghcharan, deep in the central mountains of Ghor province. I knew nobody there, and there were no hotels.

Eventually I met a young man whose family turned out to be opium smugglers. He invited me to stay in his house until I could get another onward ride over the mountains to the capital. I explained that I had no money, but Usman, my new host, explained that in Afghanistan the code of hospitality was clear: if you come upon a stranger in need, then you are bound by duty and religion to assist them. In Afghanistan, the usual Islamic practice of hospitality to guests is taken to an even greater level, where guests should be protected at all costs, even against your own neighbours.

I ended up staying at their home for a couple of nights, sitting on top of Usman's roof, looking out across to the snow-capped Hindu Kush mountains beyond. I was glad that I'd taken the risk to make it to Afghanistan, which was catching its breath after the toppling of the Taliban, and I was glad that I had put enough store in human nature, and someone else's culture, to enable me to have these experiences.

There was a blip in my faith, though. I must admit that when I was about to leave, young Usman asked me for some payment. I'd already explained that I had no money, but he really wanted a gift. I had nothing with me except the bare essentials, but I felt so bad that I had nothing to give that I ended up parting with my sleeping bag.

He seemed happy with that, but then the fear crept over me . . . what if I found myself stranded in the mountains? I might have just signed my own death warrant.

I told myself that I had barely used it since leaving Europe, and tried to put the image of freezing to death from my mind. I reminded myself that the sleeping bag was merely my comfort blanket, both physically and metaphorically. It represented fear and a guard against the unknown. But now, without anything to protect me against the elements, I was entirely at the mercy of the kindness of strangers. I had no option but to go out there and put my faith in other people. If I didn't, I would die of exposure, because now there was no plan B.

It turns out I needn't have worried. From that point on, I was taken in wherever I went and didn't need to sleep out at all until I reached the safety of India. I let go of my comfort blanket, cast my fear aside, and good things happened in return.

The Vine of the Soul

The Quechua people of Peru and dozens of other indigenous Amazonian tribes have a ceremony unlike any other.

Ayahuasca, or yagé, as it's also known, is a hallucinogenic tea made from a mix of vine, leaves and roots found only in certain parts of the Amazon rainforest. Nobody knows how the first indigenous people came across this strange potion, but it's believed to have been in use as a traditional plant medicine for thousands of years.

Shamans, or medicine men, prepare the brew by chanting poetry and singing icaras, or chants, while blowing sage smoke over the mixture in a highly formalised and ceremonial way before the tea is administered.

While the medicine has gained popularity in recent years amongst curious Westerners, who travel far and wide to partake in the ceremony, it has traditionally been reserved as a means to help people suffering from undiagnosed mental health issues. It is reputed to be extremely beneficial for those struggling with depression, anxiety and PTSD, with no known side effects.

I spoke to a Colombian shaman, who explained to me that the

medicine is useful for many physical ailments too, but only if the recipient is open-minded and willing to accept responsibility for their own problems.

It sounded strange to me – how can you accept that an illness is your own fault? It goes against everything that Western medicine states. Despite the cultures being extremely geographically distant, the Shaman's outlook on karma struck me as similar to the Buddhist way of thinking; he told me that at the root of all of life's problems was fear.

Ayahuasca works best when we overcome our fears. The medicine is ingested to the accompaniment of music, singing and chanting, sending the person into a psychedelic trance. What happens next is hard to explain, even after witnessing it first-hand. The individuals first seemed to writhe around, groaning before vomiting in the noisiest way possible. Some participants cried and screamed as if they were possessed by demons. Then, hours later, as the music became more upbeat, the crying stopped and everyone began to smile, sitting or lying still with their eyes closed. Some even stood up to dance and looked like they had found the most awesome rapture.

When I spoke to them afterwards, everyone explained that they felt like they had been on a journey to the depths of hell, having confronted their deepest fears and traumas head on. They continued to explain that through the medicine they had been cleansed of their worries and felt as if they had been reborn.

Whatever this might sound like to the Western mind, it was certainly real to those who took part. This tribal ceremony was centred around overcoming fear and facing whatever demons are hidden within. Only then can the weight of worry be lifted, and true peace found.

Heading off into the darkest Amazon to find inner peace through an hallucinogenic brew might not be everybody's cup of tea, so to speak, but it's true that only by confronting your fears can you grow and develop your true potential, in any walk of life.

Don't believe the hype

There is a big difference between real risk and perceived risk. Familiarity is a big part of how we perceive the two. If you've been doing the same kind of thing for years, you are bound to take on a new, similar risk with more gusto. Playing rugby can be dangerous, but if you started in childhood, you are conditioned to the game by the time you reach adulthood. Banging into each other at full speed would seem much more of a risk if you first started playing in your thirties, but if you're used to risking your body, then going from rugby to parachuting doesn't seem like much of a leap. In my own life, jumping off the 'cliff' as a child was scary, and so was jumping out of planes as an adult, but I was able to do it because I had increased my exposure to risk incrementally throughout my life.

There are people who think it is perfectly reasonable to avoid risk in any sense, but doing so will stunt your personal growth and likely lead to a pretty boring and uninspiring life. Risk is crucial for our development and is how we gain novel experiences and learn.

When I go travelling, I face a constant battle with other people in trying to convince them that where I am going is safe. My career as a writer and photographer has taken me across the globe to all sorts of conflict zones, war-torn regions and places that are in the news for all the wrong reasons. I often find myself in disputes with lawyers, health and safety consultants and other litigious types, who are wary of accepting the 'risk' of sending me off to somewhere like Afghanistan, Iraq or Syria.

I understand that they are just doing their job, but what they often fail to consider is that because a country is perceived as a war zone, it doesn't automatically mean that I will get shot. Conflict is usually localised – the whole of Iraq isn't at war, all at the same time. There are parts of Afghanistan that are perfectly safe for travellers to visit. The Wakhan corridor, a narrow valley in the far north-east of the country, is ethnically very different to the rest of Afghanistan. There are no Taliban there, it's been free from any fighting, and no visitor has ever been harmed in that area during the War on Terror.

I learned that, even as a soldier on 'the frontline', you spend far more time drinking tea and chatting with your friends than you do exchanging bullets with the enemy. Worry is often futile; the chances of something happening to you are really low. There's a clichéd statistic, but I think it is one worth remembering: you are far more likely to die by being trampled by a cow in a field than you are to be killed by a terrorist. You're also more likely to be injured by a toilet than struck by lightning, and you're more likely to be struck by lightning than attacked by a shark. You have higher odds of being born with eleven fingers or toes than all of the above.

The truth, as I've found out at my own expense, is that you are far more likely to be involved in a car accident while travelling than you are of being mugged, shot at or beaten up.

The media loves bad news. Moreover, the average Western person reading a newspaper doesn't care that much about how happy people are in Nigeria, or about the vibrant hipster scene in Damascus, or how literate the women are in Northern Pakistan, or the progressive nature of Iranian culture.

Don't believe everything you're told. Take the risk and see things for yourself. We must fight our stereotypes and approach ideas with an open mind, and we can only do that by embracing a certain level of risk. When we do so, we expand our field of vision and therefore our own limits – by risking more, we can fear less. I was lucky that I could go out and experience the world in order to formulate my own ideas. My conclusion is that most people are generally pretty good; even in the most 'unlikely' places, I have often found an almost overwhelming level of hospitality.

By closing your mind, you are assuming that you already have all the answers to life, which is rather arrogant when you think about it. Having an open mind doesn't mean that you don't hold any convictions, but that you are willing to be challenged on what you think is the truth, and to change your mind accordingly, as you are exposed to new places, new people and new ways of thinking. It is accepting that the process of learning, which you so embraced as a child, doesn't stop until the day that you die. Without the risk of being wrong, you will never be able to learn, and get closer to being 'right.'

All that said, risk-taking does not have to be an extreme sport; you needn't gamble all your cash on the Grand National or jump out of a plane (although I'd recommend the latter). Instead, we can get better at taking risks by making 'mistakes' less of a taboo. We should embrace risk while being prepared for mistakes, rejection or 'failure' – and I use quotation marks because is it even failure unless you give up?

We can build incremental steps in our lives to hone our capacity for risk-taking, such as visiting some place new, or switching up your regular routine – even striking up a conversation with a stranger if you are naturally shy or introverted. It may sound mundane, but these experiences will build up a tolerance and those new neuron connections will start to form. This is the explorer's mindset. We seldom explore our home cities in the way that we might when we are travelling. If you have a small choice that you need to make, try making a snap decision; leave it to the toss of a coin if you have to.

Seek things out that are *just* beyond your reach, in the foggy arena of risk-taking. Little by little you will acquaint yourself with life outside of your comfort zone. If you don't put yourself out there, vulnerable to the risk of failure, then there is no chance of progress.

Fear is a primal biological reaction. This means it is very powerful, but also that it isn't always suited to our modern lives. Like food, we tend to crave more safety than we actually need, but if we can learn to overcome our fears, we can continue to open up to the world; embracing and learning more for all of our lives. By taking risks, we learn that the world isn't as scary a place as we thought it was.

Do not underestimate the power of this realisation, as this positivity has an infectious, snowballing effect – the less we fear, the more positive we become; the more positive we become, the less we fear. And, as we're about to see in the next chapter, a positive outlook can change not only our own lives, but the world.

Annie Londonderry

4

Back Yourself With a Smile

It pays to be a winner – British Army unofficial slogan

Opportunity knocks

My grandfather served in the British Army. So did my father. After spending a few years wandering and backpacking, and loathing the prospect of a 9–5 job, I decided to follow in their footsteps. Both of them had served in the infantry – the footsloggers who carry the fight to the enemy – and it felt only natural that I would keep this family tradition going.

I passed the army's commissioning board, which meant that I could join as an officer, and command men; something I found both exciting and daunting. A young officer's training takes place at the Royal Military Academy Sandhurst and when I arrived there, aged twenty-three, I still wasn't sure where I wanted to end up. After visiting as many units as they want, to see where they might fit in, officer cadets can select two regiments to have interviews with in their final term.

It is a difficult task, as each regiment has a different role. The Cavalry might not charge about on horses anymore, but they do drive around in massive battle tanks; the Royal Military Police do everything from locking up naughty soldiers to providing special armed guards for the royal family. If you become an officer in the Logistics Corps, you could find yourself in charge of feeding front-line troops, or working out how many blankets are required in a far-off barracks.

Before going to Sandhurst, I went on a few visits to the Staffordshire Regiment, my local infantry unit, and when I was at the academy I was also interested in the Intelligence Corps. It was a very tough decision, which would impact the course of my entire life, and it weighed on me heavily during my time there. Sandhurst was hard work, and the late nights ironing my uniforms, the parades and long academic essays were only part of it. There was a lot of running, marching and being shouted at, along with field exercises, rifle ranges and blowing things up; as well as learning the more subtle arts of officership – fancy dinners, the intricacies of letter writing and knowing the Debrett's guide to etiquette inside out. It was an interesting and varied education.

I also took up boxing. Everyone had to be part of a sports club at Sandhurst, and given I wasn't into rugby or rowing, boxing seemed like a decent choice. Initially I hated the 4.30 a.m. starts, but as the fitness started to impact on me and I felt myself improving, I came to relish the early-morning alarm calls. Plus, there was the added carrot of the annual 'fight night', in which the most dedicated, committed boxers were chosen to fight, watched by the whole academy, as well as a string of generals, politicians and VIP guests.

I was incredibly proud to be chosen as one of those fighters, and on 9 November 2005, I stood in a ring surrounded by over a thousand people, face to face with Officer Cadet Mortimer. I won the fight by knocking my opponent to the ground, but I'm not retelling this story to discuss the win – I had a huge respect for my opponent, and the opportunity that came next arose simply from having stood in the arena.

After the fight, all of the boxers were invited to the Sergeants' Mess (a club of sorts, where many of our training instructors lived). This was a huge honour, because this particular mess was usually out of bounds, and the sergeants themselves were only seen in their context as authoritarian figures, all muscle, tattoos and shaved heads. The fact that we privileged few were allowed into their private domain was viewed with absolute envy by the other cadets.

The sergeants crowded round the fighters and congratulated us all on our performance, victors and defeated alike. As I finished my second pint of beer, I found another one thrust into my hand. Looking up I saw that it was Captain Truett, the Sandhurst representative of the Parachute Regiment.

'Congratulations, Wood. Which regiment are you joining?' he said sternly.

I hesitated momentarily before replying, 'I'm not sure, sir, I was looking at the Staffords, or the Int Corps.'

'Sod that,' he said. 'You should join the Paras.'

My look of surprise must have been quite apparent. The Paras were the most fearsome soldiers in the British Army. *I'm not good enough to get into the Paras*, I thought. You needed to be a muscle-bound machine to get in, surely? It wasn't even on my radar.

'Sit down,' the captain barked. I sat on one of the little stools by the bar, beer still in hand. 'The first two rounds of interviews have already been done,' he told me. 'We had over a hundred applicants and now we're down to twenty. I don't do this very often, and I won't ask you twice, but do you want an interview?'

I sobered up pretty quickly. My entire future rested in the balance and all sorts of thoughts crossed my mind: what about the Staffords, they seemed a nice enough bunch – and the Intelligence Corps, they did some interesting spy-related work? It could be a good start to my career, and would allow me to travel too. When it came to the Paras, there were a lot of unknowns. As a regiment, they are shrouded in mystery and are considered to be one of the country's – if not the world's – most elite military units. I didn't know what to expect and I didn't think I was capable of joining their ranks. The temptation was to play it safe and stick with a more achievable goal. I knew what I was getting with the Staffords.

If I agreed to an interview with the Paras, I would automatically have to turn down one of the other options, since you can only interview with two regiments, and the Paras had the toughest selection of all, just peaking the Intelligence Corps. If I turned down the Staffords and failed the board for the others, I ran the risk of not getting any of my choices – I could end up being a blanket stacker in the Logistics Corps for the rest of my career.

The captain was staring at me.

'I'd like to interview, sir,' I told him, expecting that it would take place in the coming weeks.

Nope. Captain Truett launched into a formal interview right there at the bar, grilling me about my own motivations, experiences, education and skills. After fifteen minutes he stood up, shook my hand and told me to report to his office at 6 a.m. the next morning. And that's what I did, hangover and all. I was fast-tracked into the final eight, and then at the last interview – with a panel headed by some of the most senior figures in the British Army – I was offered a place in the Parachute Regiment.

That's how I got into the Paras. It would probably have been easier to get into the Staffords, or even the Intelligence Corps, but looking back, I'm glad that I let optimism overcome my doubts about joining one of the most elite units in the world. A unique club, which opened more doors than I could have ever imagined.

Ed Stafford Walking the Amazon

Inspiration can come in many forms, and many of my own examples have been explorers from long bygone eras, but when I was in the process of leaving the army, one man's journey in particular stood out as a feat of endurance unlike anything I had heard about in the modern age.

Ed Stafford, like myself, had left the British Army as a captain some years earlier. He spent time working as a security consultant in Afghanistan and the Middle East before embarking on a walking journey that was to set the bar very high indeed. In April 2008, he set off from the southern coast of Peru with the intention of travelling the entire length of the Amazon River on foot. It was an expedition that had never been completed before. Despite his companion leaving after only a few weeks, Ed carried on, often alone, sometimes with local guides, along the course of the mighty river.

His journey was fraught with challenges; he pushed himself to the limits of human endurance, dealt with endless bureaucracy, battled deadly wildlife, narco-traffickers and the ever-present thugs working for illegal logging and mining companies. The walk took a mind-boggling 860 days (two years and four months) due to the difficult terrain.

In his book, Ed describes the emotional turmoil he went through during that period, leaving behind everything to commit to finishing a challenge he had set himself. But, and this is something I can empathise with, he was never in any doubt that he would complete it. He gave himself no option for failure, no plan B, and no safety blanket. Sometimes, only by throwing all of your eggs in one basket and risking everything, can you achieve the seemingly impossible.

Positivity breeds success

A positive mental attitude has an enormous impact on your life. Optimism will allow your mind to soar, making you open to more and more positive ideas and opportunities. It will free your mind into an ascending spiral of positive emotions, which will give you new approaches to the world and new ways of thinking. This is a huge part of being an explorer: on an expedition you need people who are constantly looking for opportunity, seeing things through a positive lens and focusing on the best. That is not to say that you don't have a sensible and healthy understanding of the negative things that might befall an adventure, but it's about hoping for the best in *spite* of that knowledge.

If you can couple optimism with enthusiasm – bringing joy, energy or genuine interest to what you do – you will go far. My thoughts resonate with those of Roald Dahl, who said:

> I began to realise how important it was to be an enthusiast in life . . . If you are interested in something, no matter what it is, go at it at full speed ahead. Embrace it with both arms, hug it, love it and above all become passionate about it. Lukewarm is no good. Hot is no good either. White hot and passionate is the only thing to be.

Enthusiasm will make people listen to you; it is contagious and energising. If you're a leader with enthusiasm, you are more likely to be followed without question, and people will call you a visionary. Enthusiasm keeps you focused and delivers results. Certainly, in my experience as a traveller, you'll have a lot more fun if you stay positive, instead of whingeing and whining about every little issue; if you see the funny side of life, you'll soon realise things aren't so bad. At the very least, *smile* – some research shows that you can even smile yourself into a better mood.

Optimism is undeniably good for our mental health, making us more resilient. This gives us the ability to cope with change or a crisis and get ourselves back to a pre-crisis state more quickly. In fact, having a positive attitude in challenging circumstances is shown to make you recover more quickly from adversity. Resilient people will translate negative feelings into positive ones, and are better at regulating their emotions. They tend to view difficult challenges as a chance for growth, rather than a threat to their wellbeing.

Studies are showing that optimism may even be good for your health. Heart transplant patients who are optimistic about surgery are making better recoveries, and optimism has been linked to a reduced risk of dying from cancer. Another recent study looked at the expectations of students about their course – how well they expected to do in exams and how optimistic they were feeling about it. It measured cell-mediated immunity, the number of immune cells that flow to an area in response to an invasion by a virus or bacteria. It found that as the participants got more optimistic, they would have a larger immune response, but when they were more pessimistic about their prospects, their immune response was slower and more delayed.

The study did not look at general outlook on life; in other words, it was not about a student's general mindset and approach, but about their specific attitude towards their course. What this means is that by being optimistic about your success in an important part of your life, you may create higher immunity against infection. It is also a crucial factor when it comes to negative health behaviours, which are things like smoking – optimists are less likely to engage in them, and more likely to be successful when they choose to quit.

Winston Churchill once said: 'Success is going from failure to failure without losing your enthusiasm.' I believe that positivity is key to success in whatever situation you find yourself. If you stay positive, then half the battle is already won.

Enthusiasm, optimism and confidence all go hand in hand to make tough circumstances bearable and give the best chance of a good outcome. Psychologists spend their whole careers looking at this, delving into how optimism increases quality of life, and perhaps more importantly, how anyone can learn to improve it for themselves. Clinical studies show that optimists are healthier, happier and more successful in life than pessimists, and this is put down to three main differences in their outlook on the world.

The first is that optimists do not personalise negative events; they view them as external, brought about by circumstances or other people. Effectively, optimists will ascribe a negative situation to causes outside of themselves, meaning that failure or struggle is not their own fault. A pessimist, on the other hand, will see their struggles and setbacks as a reflection of something that they have done themselves, or because of something internal to them. What this means is that optimists tend to be more confident, believing what they do is *in spite* of what happens around them, rather than because of it.

This confidence breeds confidence. Whether you are a soldier, a teacher, a chef or an explorer, your confidence in what you do will be picked up by the people around you. How would you feel if you went to see a doctor and they said, 'I'm not really sure what I'm doing, but I suppose I'll give it a bash.' Contrast that with, 'Don't worry, I've done this thousands of times. You're in safe hands.'

Another main difference between the optimist's and pessimist's outlook on the world is described as the pervasiveness of the situation. For a pessimist, a bad situation applies universally to all parts of life; failing in one thing will be framed as failure in life as a whole. An optimist will see this as one specific failure, and though they may feel that they are incapacitated in one area, they will have the ability to compartmentalise this. Optimists also benefit from the fact that they let positive experiences have a knock-on effect of positivity across all parts of their life, and not only that specific area.

For instance, an optimist who completed a 10K run may think to themselves, *I completed a physical challenge that I set for myself, so I know that I can succeed in that business I want to launch.*

The pessimist mindset might think, *I completed a 10K run, sure, but what's that got to do with starting a business? I'll never be able to do it.*

The third and final difference in outlook is the most crucial, and pertains to the question of permanence. If an optimist is facing a challenge, they will see this negative event as something temporary, telling themselves, *This too shall pass.* On the other hand, if you look at things through a pessimistic lens, you are assigning yourself to an immutable outcome on something that not even time can change. If things are final or viewed as a foregone conclusion, why would you bother making an effort to overcome them?

Nobunaga

There was once a powerful Japanese warrior called Oda Nobunaga who, because of his leadership prowess and great ability to motivate those around him, became known to history as the Great Unifier. There is one particular story about Nobunaga when the warlord was at the head of his army and, faced with an overwhelmingly superior enemy force, he knew he needed some divine intervention to muster his men to fight.

On the eve of battle, he visited an ancient Shinto shrine, where he prayed out loud for victory. After he finished, he went to his captains and told them that the spirit world had informed him that he should toss a coin. If it landed on heads, his army would be victorious, and if it was tails, then they should face defeat: 'Destiny holds us in its hands.'

Nobunaga prayed again in front of his troops before flipping the coin high into the air as everyone waited with bated breath. The coin landed on heads to a deep sigh of relief, and the army, emboldened by the backing of the gods, went forward and crushed the enemy. 'No one can change the hand of destiny,' remarked Nobunaga's assistant sagely, after the battle was over.

'Indeed not,' said the leader, handing him the coin that had been tossed. It was a double-headed forgery.

Never despair

Pessimism leads us into a trap where we think that everything is hopeless, and we stop looking for ways that we can influence situations positively. I have travelled to plenty of places where showing a bit of zeal and having a sense of humour has relieved me of some fairly sticky situations. I once had to go for lunch and drink moonshine with Guatemalan gangsters, in order to prove I wasn't an American spy. While in places like Afghanistan, you will often be served goat's brains and eyeballs as a delicacy by the local chief, the real test is to see if you're made of tough enough stuff and are polite enough to deserve hospitality.

One dangerous test took place the first time that I visited South Sudan, in the spring of 2011. It was a turbulent time, shortly before the country got its independence from the Arab North, and tensions were mounting as nobody knew what would happen. I had been invited on an expedition by my friend Chris Mahoney, an American who was part of the East Africa expat community. He'd been told that the UN high commissioner for South Sudan was recruiting a team of explorers to join him, and a famous water guide called Pete Meredith, to raft down the River Nile from the Ugandan border at Nimule, all the way to Juba. It was supposed to be the start of a blossoming tourism trade for this new nation.

The Sudanese civil war had raged through the 1970s and 1980s and the whole area was filled with lethal landmines and unexploded ordnance. What's more, the tribes were fighting each other again – the Dinka and the Nuer had been skirmishing over cattle and land, and the police had been up to their usual tricks of pillaging passing traders.

Despite that, I wasn't too worried, as we had a good team. There was the head of the UN mission, David Gressly, plus a bunch of other ex-pat misfits, including a a helicopter pilot, a barman, a doctor and a lion tracker.

I had some nerves, of course – crocodiles, hippos and irritable tribesmen awaited. But the only reason I'd been invited on the trip was because I was 'The Paratrooper' and with that comes a great reputation, and responsibility; so I got stuck in, helping fix up the rafts, reassuring the less experienced members of the team that things were going to be OK, and making sure that everyone's kit was strapped down properly.

For a few days we made good progress towards Juba, South Sudan's main town and soon-to-be capital city. Then, ten miles south of the outskirts of the ramshackle town, as we were relaxing and looking forward to reaching our destination and a cold beer, we were met with a nasty surprise.

There was a commotion on the river bank to our left and I could hear shouts from the bushes. I looked to the west and noticed half a dozen semi-naked tribesmen shouting and waving angrily. One was clasping a large spear that glinted in the sunlight. The men began to run along the river bank trying to catch up with us, and I couldn't help noticing that many of them were armed with AK-47 machine guns. It was clear they wanted us to stop, but even if we had wanted to, there was no chance – the current was too strong.

Crack! It sounded as if someone had taken an enormous whip and slapped it across the river.

'Shit, they're shooting at us!' shouted David.

'Paddle harder!' replied Pete.

The men on the bank fired another shot in our direction. It went over our heads, but just when I thought we were getting out of range, more tribesmen appeared, and they were directly ahead of us. There were dozens of them now – on both banks, too. Then a dugout canoe appeared from behind an island, and on second glance I made out another – not one, but two canoes, filled with angry gunmen. They were paddling hard, chasing us downstream.

'Stop paddling,' said Pete, calmly now. 'If we try to escape they will kill us. Stay composed, they're going to get us.'

Pete let go of his oar and let it drag in the river, raising his hands. We all did the same. The men in the canoes were still shouting and aiming their rifles directly at us. Eventually they caught up with us and took hold of our dinghy ropes.

As we sat there in sheer terror, now captives to these unknown assailants, we waited to see what would happen as we got dragged to the west bank.

'Get out!' one of the men screamed, with an anger that I hadn't seen before, his scarred face dripping with sweat.

We all did as we were told, clambering out of the raft with our hands still above our heads. The gunman shoved us one by one against the river bank and made us kneel down. Were they going to beat us or torture us? All sorts of horrors ran through my mind – maybe they would even execute us. This was a place where human life was cheap, and the population deeply traumatised after decades of war.

'You are mercenaries!' screamed our captor, as we were surrounded by more and more of the men. 'You are here to invade our country!'

David tried to explain that we were simply on an expedition. He protested that we were not mercenaries, that we had permission from the president himself, and that all our paperwork was in order (which it was).

The man was having none of it. 'I am the chief of police here!' he yelled. 'I am the boss here, so shut up.'

He didn't look like the chief of police, that's for sure, but who were we to argue?

'If I want to kill you all, then I will.'

What do you say to that? Nothing, is my advice. We stayed quiet. I'd been taught about how to act in this sort of situation whilst in the army, during our lessons in the art of survival and extraction from hostage situations. You are more likely to be killed if you agitate your captor, so it's important to remain calm and in no way attempt to fight or be aggressive – especially if he is being irrational and waving a loaded gun in your face.

You are also far more likely to be killed when your attacker is able to dehumanise you mentally, so it is important to try to be collected and measured, which is sometimes hard when you have a gun trained on the back of your head. To keep calm in a crisis, you have to slow down and think things through. It can seem counterintuitive but giving your next decision as much time as possible can make all the difference. Collect as much information as you can, be observant: look around and assess the situation.

In order to keep us safe, I knew that I had to remain optimistic. Focus on one small positive thing, instead of letting your mind wander into the endless possibilities of negative outcomes. If I had been standing there panicking and wondering 'what if', things might have ended very differently.

I looked over at David. He was very calm, but I could tell he was trying to do something covertly. I looked down at his hand and I saw he was grasping something. It was a satellite phone. Of course! He had somehow managed to get it out of his bag when we'd first got into trouble, but he needed to make a call, or at least alert his team we were in trouble. I had to buy him time and somehow distract the gunmen. Luckily I had just the trick – something I'd learnt on my travels.

Always carry a packet of cigarettes, even if you don't smoke. I looked at the nearest policeman, and made a gesture of smoking. He looked around to make sure the chief wasn't looking, and nodded. I took out the packet, handed him one, and took one myself. As I suspected, the chief stormed over, but before he could say anything, I held the pack open towards him and with as much confidence as I could muster, I forced a big smile, looking him hard in the eyes.

He lowered his gaze. Suddenly, I was human again. He snatched the cigarettes and took my lighter, but I kept my smile big and I could tell that he was warming to it, no matter how reluctantly.

Sometimes it is the little things that can break down barriers, and perhaps even save your life. While everyone was grinning and smoking, David managed to send out an SOS signal, and half an hour later a UN helicopter arrived filled with soldiers. We were rescued, and even allowed to carry on in our boats down river towards Juba. I wouldn't go so far as to say I made new friends, but everyone walked away unharmed, and I put that down to staying positive.

If we had all started pleading for our lives – or worse, decided that death was imminent, and so rushed our attackers – then the situation could have resulted in a bloodbath. By holding fast to the idea that an opportunity would present itself, but also that these people were not evil and intent on killing us, we came away with nothing more than some shaken nerves, and a story.

Touching the Void

In 1985, Joe Simpson and Simon Yates became the first mountaineers to climb successfully the summit of Peru's Siula Grande via the indomitable west face, but on their descent, disaster struck.

Simpson slipped and fell down an ice cliff, badly breaking his right leg. The men were already behind schedule, and had little fuel to melt snow, nor did they have any snow around them that they could drink. The pair were surrounded by ice, and Yates would have to lower Simpson down the rest of the mountain. Not only that, but they would have to do it in total darkness, and with a storm raging around them so loudly that the men couldn't even hear each other.

It is no surprise that things got worse from there, and Yates accidentally lowered Simpson off the edge of a cliff. Simpson tried to climb back up, but due to severe frostbite, he couldn't tie the required Prusik knot properly, and dropped one of the cords he needed to ascend. He was suspended over a sheer drop with one leg broken, with no way of communicating with his partner, and certainly no way of reaching him.

Above him, the weight of Simpson was beginning to pull Yates from his belay seat. Yates knew that he would plummet to his death if he didn't take action. Wracked with guilt, knowing that he was sending his friend to his death, Yates cut the cord.

Simpson fell away, but against all odds he survived the landing, and found himself deep inside a crevasse. He was suffering badly from frost-bite and his broken leg, but he was alive, and would not give in to despair. He found a route through the crevasse and onto the glacier, and from there – over the course of three days, and with barely any water or food – Simpson hopped and crawled his way across the eight kilometres to base camp. He reached it exhausted and delirious, only hours before Yates was about to depart.

The survival instinct in humans is very strong, but even so, it still beggars belief that Simpson didn't succumb to despair. From the moment he broke his leg onwards, he knew his chances of survival were next to zero, yet somehow he persevered, remaining optimistic in the face of unthinkable odds. Here we see the power of the human mind at work. Aiming for a goal with composed execution can overcome even the greatest odds.

Probability thinking vs possibility thinking

If Joe Simpson had been a 'probability thinker', he might have given up and died on the mountain. Only a focus on what's possible can pull you through an experience like that.

I find myself encountering these kinds of mindsets a lot when planning expeditions. In 2016, I started planning for a circumnavigation of the Arabian Peninsula, through thirteen Middle Eastern countries. Some of these nations have notoriously tight borders, and many others were warzones. Before we could begin to think about the probability or possibility of potential danger on the ground, we had to persuade each of these countries not only to let us in, but to grant us visas and permission to film. Many times I was told that it was unlikely we could get the requisite visas, or sufficient access to some of the countries that we wanted to include, but the members of my team and I preferred to believe it was still very much in the realm of the possible – and it was.

By engaging in 'possibility thinking', we open our perspective to entertain new ideas that were previously beyond our expectation or knowledge. In this case, the new, improved and adjusted world-view meant that I got to see the 'ground truth' in the Middle East with my own eyes, transforming my perception of the region beyond comprehension.

Now I'm not saying that one of these approaches is right, and the other is wrong – as we know from assessing risk, you need to be thinking about the probability of an incident – but I find that probability vs possibility thinking is a useful framework through which to view people's mindsets.

Of course, one person can engage in both possibility thinking *and* probability thinking, making this more of a behaviour style than a personality type. In fact, I'd say that in order to be successful, it's crucial to employ both approaches, or at least to try and hold both in mind. Once you've done your blue-sky thinking, and picturing what you would like to achieve if there were no limits, you have to return to some rational probability thinking, in order to strategise or – as is so often the case for me – plan the actual expedition.

In my experience, most people have a natural predisposition to one or the other way of thinking. It is crucial to understand this when it comes to functioning as a group, and knowing the styles you have in your team.

When I tell people about my expeditions, I can divide them into two camps based on their responses. Firstly, there are those who see them as hare-brained schemes that are bound to fail, or at least be filled with multiple challenges, which will be difficult if not impossible to overcome. They weigh up everything that could possibly go wrong, and run an analysis on whether the expedition vision is remotely probable or likely. These people are probability thinkers; they look at what *may* be. This approach of taking a reality check is logical and sensible, but because the laws of probability tell us that events and outcomes follow a cause-and-effect linearity, these people tend to be more likely to sit back and wait for an outcome.

The downfall of this way of thinking is that our ego makes us capable of doing an analysis based only on what we have done, seen or experienced in the past. Maybe the probability thinker I'm chatting to has heard only stories of drastic failed adventures, or crocodiles ripping explorers apart. It is a rear-view perspective that is based on what was possible or happened then, rather than what is possible or could happen now, or in the future.

That latter way of thinking is the realm where possibility thinkers live. For these folks, anything is possible; they look at what *can* be. Possibility thinkers are enthused by hope and have faith that things are doable. When I tell these people about an upcoming expedition, they see the boundless options and ways to enhance it; perhaps this location would be a great addition, perhaps this route might make it easier. Thinking big, bold and ambitious visions can cause our ego to panic, nervous that we are straying out of reality and into fictions that are impossible to achieve, but possibility allows you to see the opportunities and growth instead.

There is no right and wrong here. Rather, whether you are planning an expedition, or your own life, I would advise seeking out people with each mindset. The truth, as with so much in life, often lies between the two extremes.

Annie Londonderry

Possibility thinking and positive spirit lived in many explorers of days gone by, but perhaps none more so in my eyes than explorer Annie Londonderry, who was famous for being the first woman to cycle the globe. Born Annie Kopchovsky in Latvia in 1870, her parents emigrated to the United States when she was just a girl, and by her early twenties Annie was settled in Boston with a family of her own.

Annie's call to adventure came in 1894. A man had recently cycled around the world, and two gentlemen bet each other $10,000 that a woman could not do the same. It's not clear how Annie came to be the one who took up the challenge, but she did — even though she'd never ridden a bike before! The conditions of the challenge were that she must start the journey with no money, and that she must finish it in under fifteen months.

Annie assembled five hundred or so people to see her off and used this as a publicity opportunity. As well as fundraising for the whole trip on her own, the savvy young woman had arranged for sponsorship from a water company called Londonderry Lithia. They would pay her to put an advertising plaque for their brand on her bicycle, as well as change her surname to Londonderry. Annie got a representative of the company to hand over the $100 in front of the crowds, and this gained her a lot of applause and attention.

In France, Annie's bike was promptly confiscated by customs and the French press wrote insulting stories about her unladylike clothing. Undeterred, Annie continued on and made it to Egypt, Jerusalem, the port of Aden in Yemen, then across to Colombo, Singapore, Hong Kong, Korea, Russia and Japan before taking a boat back to America. Annie collected signatures from the American ambassador in each country, and made it back to America fourteen days under the allotted time. Apparently, she never got on a bicycle ever again!

Annie Londonderry was not without controversy, though. She certainly knew how to bend the rules, hopping on the odd train or ferry and taking her bike with her, on the basis that the rules stipulated she travel 'with' her bicycle. As well as this liberal approach to transport, it turned out that the wager was entirely fictitious – there were no two gentlemen betters. While there are some slightly questionable elements to the trip, there is no doubt that Annie worked hard and took things into her own hands, creating her own luck.

She was the ultimate hustler, putting her bike out to tender for advertising space in return for cash to keep herself going as she rode through towns. At one stage even her clothing was covered top to toe in adverts, a mobile billboard. She developed her own line of merchandise, selling autographs and photos of herself, as well as giving cycling demos.

Annie was a phenomenal saleswoman and a captivating storyteller: she knew exactly which juicy and gory details of her adventures to share with her growing audience.

She was astute and talented at making the most of the media and a master of public relations. On the second half of her journey, she earned extra pocket money by delivering lectures about her adventures. Resourceful and enterprising, her enthusiasm and possibility thinking made her arguably the world's first international cycling star.

Discovering optimism

Optimism and pessimism, like fear and the fight-or-flight reflex, operate on subconscious levels, but we can use our conscious minds to rewire these reflexes. For example, we can visualise the best possible versions of ourselves, and describe them out loud to ourselves using mantras, or other verbal or written reminders. This will change our subconscious from thinking negatively about our flaws, to thinking positively about our potential.

We all have a negative internal monologue – the voice in our head that tells us something can't be done. Because we have a tendency to be a risk-averse species, it is easy to think that this voice is rational, but in truth it's hopelessly not the case, especially in the modern world we live in. We can become more positive by habitually pushing back against this negativity with rational, optimistic arguments; counterbalancing worst-case scenarios with best-case predictions, in order to identify more realistic middle-grounds.

Above all, we can completely reconfigure how we approach the concepts of disappointment and failure. Our inner pessimists make mountains out of molehills wherever setbacks are concerned, and perceive these shortcomings or blockers as fixed and permanent. But the reality is that very, very few things in life are beyond our control, and if we come to view setbacks as learning opportunities, then life immediately becomes more positive. We are always either succeeding, or learning.

One way that you can increase or better learn optimism is by practising gratitude. Think of a time in your life when you've felt or expressed extremely heartfelt thanks for something that someone else has done for you, or said to you. Of course, this feeling of thanks can also be directed to yourself, not just another. It's a powerful feeling. Every day, write down something that you're thankful to yourself for, and then another thing for which you're grateful that came from someone else. It doesn't have to be a physical item. Perhaps you are grateful for the kind ear that somebody gave you, or the conversation with a stranger on the train. It helps us to see the good qualities in others, adopting and picking up behaviour that we admire.

Studies show that more grateful people tend to have a stronger sense of belonging and are less likely to suffer from stress or depression. Practising gratitude breeds energy and enthusiasm, which an explorer would be lost without.

As in Chapter Three, where we looked at what scared us, the more we can push ourselves outside our own comfort zones, and face our fears, the more optimistic we will become. Fear is deeply rooted in the unfamiliar: it is when doing unusual things like flying in a plane that we tend to overestimate the dangers, compared to everyday things like travelling by car. Becoming familiar with the unusual, by expanding our comfort zones, will make us more optimistic over time. There is a compounding effect to it when we consistently show ourselves that we can come through situations that we perceived as difficult or dangerous.

Optimism in the face of adversity and uncertainty are the building blocks of good preparation. We can't control everything in our lives, and there will be times when things do go wrong, but what we can *always* control is the way that we react to those hardships. It is down to us, and us alone, to look for the possible over the probable. To see obstacles as opportunities. And to recognise that today's disaster is tomorrow's laughter with friends. With a positive mindset, you may still encounter setbacks in your life, but you will never lose.

A Polynesian Chief

5

Be Ready For Anything

By failing to prepare, you are preparing to fail – Benjamin Franklin

Public speaking

Like a lot of people, the thought of public speaking was something that crippled me with nerves. I'd had to stand up in front of people in the army, of course, but there is something about the uniform and rank that offers you protection. You're delivering a lesson or mission given to you by the army, not putting your own life on display for others to judge. Besides, it's not like my soldiers could choose to stand up and walk out if I was boring them!

My first offer to give a speech came out of the blue, but apparently setting up an adventure travel club and leading a few small expeditions warranted my invitation down to the local Rotary Club to speak to a handful of elderly gentlemen.

Before I went on stage, I was genuinely terrified. I got a flood of self-doubt and panic, convinced that the entire thing was going to be a mess. My heart beat faster in spite of my best efforts to slow it down. Comedians talk about 'dying on stage', and as much as my rational mind could tell my sweaty, shaking body to calm down, my flight reaction of feeling cornered, with all of those faces looking up at me, made my body involuntarily panic. My legs had a slight wobble, which I probably didn't manage to conceal successfully, and my hands were clammy. I was told once that this feeling of stage fright evolved from an ancestral tradition, whereby our forefathers were summoned in front of the tribal elders, usually in order to defend their actions when some wrongdoing had occurred.

I'd been fine for days beforehand, comfortable that I knew what I was going to say and that I was capable of communicating something coherent. Why then, when I was about to go on stage, did I suddenly feel that all of my lines were going to slip totally from my mind? Perhaps it was because I felt the stakes were higher here. These were my elders, and money was at stake – I was fundraising for a charity and some of the audience were local councillors and businessmen, and one was a former brigadier.

This is the part where I tell you that I overcame my fears, and delivered a Churchillian speech, right?

If only. The talk was a complete shambles. The projector stopped working, I forgot my lines, and nobody laughed at my jokes. It was a disaster, but when it was over, I realised that I was still standing and alive – despite what my body had been telling me. Just as I was consoling myself that hopefully they were all hard of hearing anyway, the chap who had invited me to talk came bowling over.

'Levison,' he said. I stood to attention, which probably says something about how detached my mind and body were; at that point I was working from muscle memory. 'That was fine as far as these things go, young man. An interesting story about Afghanistan, but you can definitely do better.'

He proceeded to provide a lengthy critique on how I might best improve. It was pretty embarrassing, but now I look back at it and laugh. In truth, I had probably not practised or prepared enough. I'd been too worried about what people thought, and not spent enough time simply rehearsing the content and making sure the projector worked.

Anyway, it was over and done with. The audience coughed up a couple of grand for the charity, and I escaped with nothing but a bruised ego and a few valuable lessons learned.

For me, stage fright has now pretty much abated. To be honest, I managed to overcome it by making one pretty straightforward change: I learnt to tell myself that I was excited, not nervous, and to see every potential moment of fear as an opportunity to grow and learn. To re-channel the feelings and tell myself that it was simply a healthy dose of anticipation and enthusiasm about going on stage.

As I got ready at the Rotary Club, I had been thinking about all the things that could possibly go wrong. In a disastrous downwards spiral of negative reinforcement, I thought to myself, *I'm just being realistic.* I hadn't practised enough, so this was a reasonable expectation, but if you are negative about something, I'd say it's doomed from the start.

Instead, if you adopt the mindset that you are ready for anything, and anticipate that anything can happen, then you'll never be taken on the back foot. That's not to say the fear will entirely disappear, but you remind yourself to keep things in perspective. Even if you don't feel that way, if you project it, others will buy into it, and this will reinforce your belief – 'fake it till you make it' – but nine years after my Rotary Club debacle, I found myself in another public speaking situation that was even more terrifying.

I am fortunate enough now to call myself a 'high profile supporter' for the charity UNICEF. I get to visit their amazing projects around the world and help promote a really important cause, for which I feel incredibly privileged. One of the perks of the job is to be invited to the occasional VIP reception. One such event was the England vs India World Cup cricket match at Edgbaston in the summer of 2019, for which UNICEF was the beneficiary charity. I was invited to watch the match from the press box alongside some of the cricket greats, including Sachin Tendulkar.

Over lunch, one of the UNICEF press team asked if I wouldn't mind appearing on a pre-recorded radio or TV interview to talk briefly about the great work being done by UNICEF, and of course, I agreed. After doing a couple of brief chats, which were very straightforward and about my own visits to UNICEF projects, I was asked if I would do one more.

'No problem.'

'It's live?'

'Sure, no problem.' I had done bits on TV before, so I wasn't worried.

I was led out of the press room and down the stairs; I figured to a position with a more interesting backdrop for television.

I was half-right.

Before I knew it, I was being led out in front of 25,000 people. There was a box in the centre of the pitch, and there might as well have been an executioner standing beside it – I felt sick, and all of a sudden my knees were wobbly. It turns out that you can be a para-trooper and go to war, but the eyes of a stadium upon you is far more daunting than Taliban bullets. If I had been hit on the battlefield, I'd be a hero. If I messed this up, I'd just be a prat.

As I was handed a microphone, I tried not to think about the millions of people watching my address around the world.

My saving grace was that for the last ten years I had been practising my public speaking, and that practice had bred confidence. It didn't take away the wave of fear, but it allowed me to see over it. And so, I puffed out my chest, took a deep breath and waffled something about what a great atmosphere it was, welcomed the Indian crowds, and told everyone to dig deep and support UNICEF. It was a roaring success and I walked off the box exhilarated.

There was more to come. As I was walking back up the stairs of the pavilion to the press box, I was rushed by Indian fans who wanted me to sign their bats and T-shirts, which of course I did heartily. It was only when one shamefaced boy asked me what team I played for that I realised not a single one of them had a clue who I was. While this was a little humbling, it was also a good reminder that if I *had* fluffed the whole thing up, nobody would have given a stuff but me. A significant moment in our own lives can be nothing more than a few seconds of distraction in someone else's – if that – and soon forgotten. It is us who will live with our performances, and us who must prepare.

Body Armour

According to traditional belief, the Tahitian god Ta'aroa, 'The Severer', created the world by separating it into two parts: the po, the world of darkness, night and death, and the ao, the world we know of light, daytime and life. This Tahitian view of the universe having come into being through an act of creative separation, or something like it, underpins the belief systems of native people across Polynesia, and continued to influence Polynesian culture even after Christianity spread throughout the region.

In this belief system, the two worlds are in perpetual threat of merging together once again, destroying the universe as we know it. Particularly vulnerable to this spiritual, supernatural threat are the backs of people – those parts of the body that we can't see – which represent an ever-present but invisible threat, like the po itself. That's why, in these cultures, it was forbidden to walk behind certain important, high-status individuals.

Men, especially adolescents on the verge of full manhood, were particularly vulnerable to the dangers of the po lurking unseen behind them, particularly where warfare and sex (because of the semi-sacred status of women) were concerned. In war especially, death in battle was viewed as a result of spiritual shortcomings as much as anything physical or martial; so before taking part in either love or war, adolescent Polynesian men needed to prepare themselves properly in order to gain spiritual protection.

Part of this protection took the form of tattoos. Tattooing in Polynesian culture creates a protective layer covering the body, warding off the spiritual dangers of the other-world. In Samoa and Tonga, this was typically represented by wrapping the body in a protective shell, emblematic of the natural armour of animals such as tortoises, or the cloaking wings of a flying fox.

In the Marquesas, tattoos often feature secondary faces, either representing protective Gods, or symbolic extra eyes to watch for the unseen cosmic dangers behind the wearer's back. These might often depict a skull, denoting the spirit of a dead chief thirsty for vengeance, protecting a warrior's back in battle.

Tattooing was an essential part of the Polynesian initiation into full manhood for young warriors, who would not even think about going into battle until they were properly prepared, physically and spiritually.

The same applies to your mind: success is an art form. Prepare yourself by polishing your own mental armour and victory will be yours.

There is no such thing as luck . . .

The motto of the Parachute Regiment is *Utrinque Paratus,* a Latin phrase that translates to 'Ready for Anything'. I learnt in the Paras that being prepared is the key to success, and it is something that has served me well both in my career in the army and beyond.

The Paras were formed in the Second World War to act as shock troops, light on their feet and ready to jump, literally, into any situation. They have been used in that same capacity since they were formed in 1942 and they are trained to the highest standard in fitness, navigation, close-quarter combat and survival in the harshest environments. But the main reason that Paratroopers have been used by conventional armies around the world over the last seventy-plus years is because of their exceptional resourcefulness – their ability to deal with rapidly changing circumstances, to fight on with limited kit and low ammunition, and find their way out of trouble.

Being 'ready for anything' is about having the right mindset; it's about preparing for the worst but hoping for the best. Paratroopers spend countless weeks and months on exercise in the pouring rain, going over the same weapon-handling drills, practising their marksmanship, getting their muscle memory in order, and making sure they are fitter than everyone else. That is because they stand by an old soldier's adage, often attributed to that most tenacious of Frenchmen, Napoleon Bonaparte: 'Train hard, fight easy.'

I also believe that you make your own luck in the world, and you do so by being ready for anything. Take the example of a professional boxer. In training, they constantly practise the same manoeuvre over and over again, to the point where their brain and body are so attuned to the process that it becomes second nature to them – even automatic – and once they have mastered that task, they can focus on more important things like winning the championship belt.

Similarly, when a soldier ducks down to avoid a bullet by mere inches, they may well be 'lucky', but perhaps that luck has been engendered by years of training, of becoming aware of their surroundings and in tune with their environment, and having the instinctiveness to know when to duck. As my sergeant explained to new recruits about how he had survived so many near-misses: 'The more I practise, the luckier I get!'

In my experience, lucky people are also busy people. They are doers, not pontificators. They are busy putting themselves out there, which helps them to create more opportunity for themselves through moving in new circles, or meeting new people. They are adept at networking and seeking out chances for success, and finding themselves in the right place at the right time. They tend to be great at observing the seed of an idea or an opportunity and seizing it; they are ready for anything.

There are two possible mindsets we can take to challenges or adversity; an internal locus of control, or an external locus of control. The latter thinks, *I am beholden to events, the world happens to me*; the former thinks, *I can shape events, I happen to the world*. People who adopt the former are the doers, the ones who believe you make your own luck and, simply by subscribing to this belief, they probably do.

We create success or failure through our thoughts. Unlucky people, on the other hand, know how to talk themselves out of an opportunity, to find a reason not to do something, as opposed to chancing their arm. They forget that, first and foremost, the prerequisite for luck is action, and making stuff happen.

As Tennessee Williams, the American playwright, put it: 'Luck is believing you're lucky.' Lucky people are convinced that their future will hold great things. Much like the power of positivity and optimism, these expectations can become self-fulfilling prophecies, where they believe that things will go well, and soon they do. This world-view is all part of the explorer's mindset, and as it gets repeatedly reinforced, it allows lucky people to persevere even when they are knocked back. Like a super power, they have a tendency to turn misfortune to their advantage.

It is all about mindset; those who have a resilient attitude to setbacks are inevitably stronger, more opportunistic and ultimately luckier. This subconscious psychological work allows them to reframe things — a set-back or a negative event is immediately construed as 'that could have gone worse', or 'at least this didn't happen'. What might be viewed by a pessimist as bad luck is reimagined as not so bad after all. Perhaps above all else, lucky people get good at listening to their instincts, and this ability to trust their intuition makes them effective decision makers.

Practice makes perfect, or if there's no such thing as perfect, at least it makes us fortunate. It changes our mindset from one where 'bad things happen to me' to one where 'I happen to bad things'. We become more resilient to adversity, because we believe that we can turn the tide in our favour if we keep working, practising and preparing ourselves for challenges. This is all part of the art of exploration: being tirelessly conscientious, and turning yourself into a lucky person through diligent, thorough preparation and practice.

The Six Ps

Expeditions rarely go to plan. You spend months preparing and lining everything up, but even with the most stringent contingencies in place, something will go the way you didn't expect. Of course, the thing to do in those situations is imagine just how badly it would or could have gone if you *hadn't* done all the legwork and prep before-hand. That's not to say it all has to be doom and gloom – you can hope for the best, as long as you prepare for the worst.

This is an adage that the army lives by, usually uttered in barracks by grizzled sergeants: 'Prior Planning Prevents Piss Poor Performance.' The importance of the Six Ps was a lesson I learned the hard way, time and time again.

The rain in Spain

In January 2005, as was often the case in my early twenties, I found myself eager for adventure but too poor to afford it. I had recently got back from a long trip hitchhiking to India, where I had spent all my meagre savings. I was staying at my parents' house in Stoke-on-Trent, waiting to join the army and go to Sandhurst to begin my military career, but that wasn't until May. I tried working for a bit, doing odd jobs, manual labour, and working as a hired hand filling up vending machines. It was enough to earn a few hundred quid to see me through the next months, and into the profession I had waited so long to join.

Stacking vending machines didn't give me quite the same thrill that travelling to India had done, and by the end of the first month I had itchy feet. Even though I had only a pittance to my name, I decided that I'd rather spend my time broke and on the road than being bored at home.

I've always had a love of maps, and I was soon poring over them. Given my limited resources, Europe would be my destination. I reckoned I could get away with a week-long trip somewhere if I camped out and lived on one or two meals a day. Outside the home in Stoke it was driving rain and windy, with temperatures hovering around freezing. I needed to go somewhere warm.

Spain suddenly jumped off the page. I had been reading *Don Quixote* in between work shifts, and fancied the romantic idea of exploring the Spanish countryside. I imagined its olive groves, crystalline beaches and sultry ladies. What's more, my friend Tim from university had invited me to visit him in Gibraltar. So, a trip to Iberia it was.

Now, how to get there? I checked flights to Madrid, and managed to get a very good deal on a low cost airline for the first week of February. I thought to myself, *what the heck?* I could make it from

Madrid to Gibraltar in four or five days easily enough if I relied on local help, but since I'd just finished a big hitchhiking trip I wanted something different, and I couldn't afford the trains and buses.

In a moment of inspiration, I decided that I would cycle. I didn't really have the money to rent one – and certainly not buy one – but I had one in the garage from when I was thirteen. That would do!

I packed a small bag, then pulled the bike apart and wrapped the whole thing in cardboard so that I could check it in as my luggage. I liked the idea of a spontaneous adventure, so I did absolutely no route planning. This was 2005, before Google Maps and TripAdvisor, so I would be entirely reliant on my trusty Lonely Planet guidebook and local directions.

The next week I touched down in Madrid, alone and excited. Unpacking my bike, I got it ready for its first road trip, and off we went, heading due south. For the next hundred miles or so, I weaved amongst and hill-top towns of the Spanish plains. I passed by Toledo, with its magnificent palace and dramatic bridges, and the lovely villages of Southern Castile. At first it was fun. A lot of fun. I was feeling pretty bloody proud of myself, to be honest.

And then it began to rain.

And rain.

And rain.

Whoever said 'The rain in Spain falls mainly on the plain' is a bloody liar. It turns out that the rain in Spain falls everywhere, especially in February. My grand plan to sleep rough fell at the first hurdle, and two nights of sleeping in a soaking wet bivvy bag at the side of the road was quite enough. The further south I got, the more miserable the situation became. I had massively underestimated the distances and even more overestimated my own cycling proficiency. In a bid to keep to schedule, I stayed mainly on the highways, which was both dull and dangerous, particularly when lorries flew by, beeping their horns as hailstones as large as bricks pounded the tarmac.

On the third day of cycling, I reached the industrial town of Ciudad Real late in the evening. I was hoping to find a hostel dorm bed for the night, even though I could barely afford to eat, but my guidebook advice on Ciudad Real was bleak: 'It's a gritty Spanish working town where tourists rarely venture, and there's not enough here to warrant a detour off the main highway.'

Brilliant.

But it was already dark, and I couldn't face another night in a ditch, so I peddled hard to reach the town centre, where I might at least be able to sleep on a bench in the train station.

I parked my bike outside, found a suitable dry corner and did exactly that. My sleeping bag was still damp from the night before, but it was better than being in the open. I took a whiff of myself, and I smelled pretty bad – it was probably for the best that I wasn't in a shared dorm room after all.

Just as I was falling asleep, I received a sharp dig to my ribs. Looking up, I saw a stern Spanish policeman, who was holding a baton and yelling loudly. I didn't know much Spanish then, but it definitely did not sound like he was asking me how my day was going.

I was at a loss as to how to reply, so I did the ridiculous hand motion that we associate with cycling. Maybe this means something very different in Spain, because a few seconds later I was bustled into handcuffs and dragged to the police car; my protestations about leaving my bicycle behind falling on deaf ears. Ten minutes later, I found myself at a strange building on the outskirts of town that didn't look like a police station apart from the bars on the windows.

Feeling rather sorry for myself, I was led down a dingy corridor to a room with a single bed, a washbasin in the corner, and a picture of Mary and the baby Jesus on the wall. I shrugged my shoulders as the copper pushed me into the room and locked the door behind me. I may have been locked up, but I did have a warm bed for the night.

The next morning I awoke at 6 a.m. to a knocking on the door, and a little lady beckoned me to follow her. I was led into a communal kitchen area with a large wooden dining table, still utterly confused, where the lady gestured that I sit, then a bowl of porridge was placed in front of me. I looked up from my breakfast as bearded, hairy men began to take their seats. Soon I was not the only one eating porridge, and then it hit me.

I was in a homeless shelter.

After breakfast, I was unceremoniously booted out and told to take my vagabonding elsewhere. It was a long, miserable walk back to the station to collect my bike. It was still there, but after that night I didn't do any more cycling, instead I hitched a lift to the south coast, and spent a few days in Gibraltar with my mate Tim. He found the whole episode very amusing, of course. For me, it was a lesson in humility, and the importance of preparation.

I also decided that cycling was not my thing.

Sir John Franklin

In May 1845, Sir John Franklin led a Royal Navy expedition to discover the Northwest Passage; a stretch of water north of Canada linking the Atlantic and Pacific oceans. He had two ships and 129 men under his command.

According to the subsequently maligned Arctic explorer John Rae, who later discovered the remains of the ill-fated expedition, this was Franklin's first mistake. The local Inuit peoples, who Rae spent many years living with and learning from in order to survive the harsh Polar conditions, never travelled in groups of more than ten or twelve, as the land can't support any more.

The ships were laden with useless, unnecessary burdens: a 1,200-book library, sophisticated cameras, silver cutlery and candlesticks, and even a piano with reams of accompanying sheet music. Also aboard were three years' worth of tinned food supplies – purchased short-notice from a cut-rate supplier, whose hasty soldering process allowed lead to seep into the food.

No one knows exactly what happened to the expedition, but Rae's conversations with Inuit groups near King William Island, as well as evidence on recovered bones, suggests that around forty of the crew died of starvation after resorting to cannibalism on the island. A combination of malnutrition, extreme conditions, lead poisoning and scurvy meant that none of the expedition's members survived. A note later found on Beechey Island revealed that Sir John Franklin had died there on 11 June 1847.

Bad luck contributed towards this tragedy, but poor preparation meant that the crew were especially unable to respond when misfortune struck. The mission was, in effect, doomed before it had even begun.

Hatch a plan

I loved my time in the army for the most part. It had its highs and lows of course, but after five years in the Paras I was getting a bit restless.

In the autumn of 2009, I decided to apply for Special Forces selection, but made the error of going horse riding in Mexico with my girlfriend's family the week before I was due to begin the notoriously difficult process. That was not a good idea, because I had as much luck horse riding in Mexico as I did cycling in Spain. I came off the horse, fracturing my ankle, and with my injuries, there was no way I was going to be able to run up and down the Welsh mountains with a rucksack on my back.

Unfortunately, all of the posts in my regiment were filled. I hadn't been assigned one, because I was due to go on selection, and the Parachute Regiment is not the kind of place where you hold something back for someone in case they fail. I went to speak with personnel to see what I might be able to do for the next year, and discovered to my surprise that my commission was nearly up. The Queen's commission gives an officer their authority to serve, and is, in effect, one's contract. I saw this as a sign, and rather than seeking to extend my commission, I secured myself a few weeks' gardening leave and mentally began to pack my bags.

Once I broke the news that I was leaving the army, the question everyone wanted to ask was, 'What are you going to do next?'

The truth was that I had no idea. Not a specific one, at least. I knew what I liked and I what I didn't, but rather than rush into a job that I might end up hating, I decided to take stock and make a plan instead – as you can see, my time in the army had taught me the value of preparation. Amazing how that lesson can be learned when you have an angry sergeant major to 'positively reinforce' it for you.

I was very fortunate that a lot had happened in my life up to this point, and I wanted to reflect on that so I could plan for the future. When the time came to do so, I sat down with a notebook and pen and used a technique called the combat estimate. In the military, whether you're a platoon commander leading thirty men to assault a bunker position, or Napoleon himself advancing on Moscow, you use this process. It's a plan, refined over centuries of warfare, designed to give commanders a clear picture of the options available to them.

There are many versions of this that leaders have used throughout history, but the British Army has simplified it to seven questions:

1. *What is the enemy doing and why?*
2. *What have I been told to do and why?*
3. *What effects do I want to have on the enemy and what direction must I give to develop my plan?*
4. *Where can I best accomplish each action/effect?*
5. *What resources do I need to accomplish each action/effect?*
6. *When and where do the actions take place in relation to each other?*
7. *What control measures do I need to impose?*

My days of planning operations were over, but what had worked in combat could work for me in life – after all, we're in a war against time, aren't we? A battle to fulfill our potential, to experience the world and to leave it a better place than we found it.

That being the case, I set about changing the terminology to fit my new goals, and adapted the formula into five questions for life planning. For each question, I gave myself a framework and method for how to answer it:

1. What is the current situation and how does it affect me?
   Method: Conduct a SWOT analysis. What are you good at? What do you enjoy? What will earn you enough cash to support your goals?

2. What do I want my mission to be and why? = mission statement, main effort, timeframe.
   Method: Write a mission statement, and set a REALISTIC timeframe.

3. What effects do I need to achieve and what direction must I give to develop the plan?
   Method: Write out my goals, and then the most important steps to achieving them – work backwards!

4. What resources do I need to accomplish what I want to do? = asset requirements.
   Method: List asset requirements – money, equipment, team members, skills, etc.

5. Where and when do the actions take place in relation to each other?
   Method: Synchronisation matrix – how do all these actions fit together, what are my options? A spreadsheet or a mind map.

If you want to find out more about the planning process, you can find templates online.

I can't recommend them enough, as the more questions I answered, the more the plan for the next stage of my life came together. I began to see a loose structure form on the paper in front of me. I made a list of all the things I was good at, and all the things I enjoyed, and suddenly these pie-in-the-sky ideas began to morph into a set of potential goals and challenges.

Thanks to that framework I knew what I wanted, and therefore my radar was set to picking up the right opportunities. I began talking about my ideas with a friend, Tom Bodkin, who was also in the process of leaving the Paras. He was keen on travel, loved winter sports and had a similar mindset to me. Because we'd both worked out what we wanted from the next stage of our lives, it was easy to make the decision to launch our own expedition company, and Secret Compass was born.

We were both experienced in leading expeditions, and more to the point, we both loved doing it. The idea was to run cool trips to far-flung destinations, but we weren't any more specific than that. Who our clients would be, and where we would take them, was something that we needed to determine, and so together we did a second combat estimate, but this time on Secret Compass as a business specifically.

I believe this is a very important step. It is not enough to conduct only one appraisal of your life and goals and be done with it. Keep refining. Keep honing down. By doing this, Tom and I decided that we'd use our military experience and take people to post-conflict zones, or areas that other companies were too scared to go. We pioneered a series of world-first expeditions, beginning with a horse-riding trek across the most remote valley in Afghanistan, and followed up with mountain climbing on the Iraq–Iran border. (We even took the first-ever bloke skiing in Kurdistan, which involved dodging a minefield – luckily, there was a lot of snow!)

It was all done with military precision to the highest standard of service and risk assessment, and the hard work and preparation paid off. In less than three years, Secret Compass became a global market leader.

I organised and led expeditions all around the world, from the jungles of Madagascar to the frozen forests of the Arctic. It wasn't long before I started getting phone calls from big news agencies and broadcasters asking if I would help them to get access to the places we were taking tourists. We became the experts in the field, and were described by many as 'an overnight success'. Tom and I laughed when we heard that, but we were far from the only people out there whose results of dogged preparation had been described that way.

Jonny Wilkinson

Ever since he kicked England to a win in the 2003 Rugby World Cup final, Jonny Wilkinson has become a household name in the UK. He is recognised as one of the greatest players of all time, and became one of rugby union's first millionaires.

What a lot of people don't know, or at least don't realise about 'Wilko', is that none of this came to him naturally. He wasn't born with an ability to play rugby. In fact, as a teenager he was rejected at county trials for the Surrey Under-15s team. More gifted athletes cruised onto the team, but where are they now?

Wilkinson did not have the natural gifts at that age, but the reason his name will be remembered is because of his near religious commitment to practice.

Throughout his career, he wouldn't leave a training session without having successfully kicked six drop-goals in a row, imagining each of them to be the deciding kick in a World Cup final. As he explained in an interview to ESPN, a few months before the 2003 World Cup:

'One afternoon it took me about an hour and a half . . . the balls

were blowing here and there in the wind. I got tired and angry that day too, because I know some days I can aim at a dot and actually hit it over and over again. I was constantly getting the fourth and fifth, but missing the sixth . . . Great players and kickers are great because they've done fantastic amounts of great practice. Everything you've done since you started is there in the bank to be drawn upon.'

When the big moment finally came for him, as he'd practised over and over again for years on end, Jonny was well-prepared. He had set his standards high, and whatever the external conditions, he'd held himself accountable for hitting them. As the saying goes – train hard, fight easy.

Being prepared

There is no such thing as an overnight success; it took diligent preparation for Tom and I to get Secret Compass up to where it was. Behind the scenes there had been a steady slog – hours and hours put in to get there.

Whatever it is you've got earmarked for your next venture in the world, however big or small, start by making a plan of action. This might involve something like the combat estimate, or if it's a more personal thing such as learning a new skill, make a plan of the resources you will need to do it. For sake of example, let's say you want to get better at cooking.

First among these resources is always time: when are you going to practise your cooking? Put space in your diary (if you don't use one, now's a great time to start – your phone will have one built in) on a weekly or daily recurring basis, and view that time as sacred. Thursday evening is cooking night – no ifs, no buts. Then, once you've secured yourself some time, think about the other things you'll need: materials, equipment, and instructions (i.e., ingredients, kitchen utensils and cookery books).

Identify any gaps, and get these filled before you start – this is really important, as the frustration of finding you're missing something crucial can kill the positive momentum that builds in the early stages of any new project. Plan what you're going to do in each session you've booked. This Thursday, Pad Thai; next week, chocolate brownies.

Of course, the bigger the project, the more planning will be needed. Think about each step of the process in detail, and question if there's anything you might have taken for granted that could cause problems if it goes wrong, or missing. This is especially helpful in working contexts. Big pitch to a client tomorrow? Read through your deck tonight – if you can, ask someone close to you to watch you rehearse. Whatever your chosen skill or area, practise and practise until you can do something in your sleep, and then you will not fail.

You must plan, but always remember to stay flexible; there is no use sticking stubbornly to a plan that is not serving you. Life is not neatly pre-packaged with clear instructions for use. You have to allow a space for possibility to flourish – you can't just live your life according to a five-year plan or a projection of what you want it to look like. By doing that, you will shut down true opportunity when it appears. Embrace those opportunities, anticipate them and pull them towards you. That's why it is important to set goals, and come up with options on how to get there.

By being goal oriented, you give yourself the best chance of understanding the bigger picture and not being distracted by easy wins; and by educating yourself in the different levels of detail required to succeed in any task, you can try your hand at a bit of everything, not only what you are good at.

On an expedition, preparation goes a long way in achieving the aim. The same goes if you are starting a business or writing a book: you may not have any experience or deep love of accountancy, but you – or someone you know – will have to do the job, so it makes sense at least to begin to understand what is required. Using positive thinking, figure out where your strengths lie.

Once you are prepared for anything, you can start focusing on the things that interest you the most, and then, for all the other stuff – DELEGATE. Even the most accomplished people cannot succeed alone. Expeditions, like all projects, need a solid team, and, as the next chapter explores, solid teamwork.

Buzz Aldrin on the moon, with Neil Armstrong

6

Build Your Tribe

In order to achieve great things, you must surround yourself with great people – anonymous

Strength in diversity

In 2002, I was studying history at university in Nottingham. I was in my second year and I'd made a good bunch of friends. There were people from my halls of residence, my course and a few randoms I'd met in the library. But the people I liked best were the crowd I met at the University Officer Training Corps.

The UOTC was technically a branch of the army, aimed at students to give them insight into military life. We were non-deployable, so we couldn't be sent off to fight a war, but we did get to go on exercises, shoot rifles and do training courses; everything from parachuting to kayaking and hill-walking in Wales. I loved it, and I ended up living with a bunch of fellow cadets in my final year at uni.

Every Wednesday night, after we'd finished our military training, we'd have a few beers in the mess bar before going out into town to go drinking. We'd usually end up in one of the regular student nightclubs, but on one particular night our drunken foray happened to coincide with the annual hustings for 'Mr and Miss Nottingham'.

It was a talent contest of sorts, where the applicants had put themselves forward a few weeks before and been selected on their looks, banter and general extravagance. The entire student population then gathered and crammed itself into the Palais nightclub to see the final competition. The three male and female finalists were to parade themselves on stage, show off whatever talent they had, and whoever got the biggest cheers from the drunk crowds would be the winning boy and girl.

The girls were up first, three stunners who gyrated to the music and downed shots of tequila. One of them did a cartwheel, while another did an impossibly low splits, much to all the boys' delight. The one with the biggest boobs won.

Then it was the boys' turn. One of them was a hulking rugby player, another a skinny but amusing comedian who wooed the girls with his acerbic wit, while the last one was a lad I recognised. He was one of my friend's housemates, a philosophy student who went by the name of Ash Bhardwaj. I remembered seeing him out and about, but had always found him too loud, too gregarious and brash. I hoped he wouldn't win.

The rugby player pranced around first, flexing his muscles and banging out hand-clap press-ups. The skinny lad couldn't compete with that, so he had his friend pass him a 'dirty yard' – a metre-long glass receptacle filled with two and a half pints of various alcoholic drinks, including an obscene amount of spirits – which he proceeded to down in one, before promptly vomiting the whole lot straight down into the faces of those unlucky enough to be in the front row. The crowds went wild.

Then it was Ash's turn. He pointed at the DJ and gave a thumbs up. The theme tune from *Baywatch* began, blaring out its familiar beats. Ash stood stone still at first, before ripping his shirt from his chest. Then he whipped off his shoes and jeans, whirling the lot around his head. Swiftly after that, off came his underpants. He was now fully naked in front of two thousand people. Finally, to top off the show, he lobbed his wardrobe straight into the crowd, with one of his shoes cracking me directly in the face.

What an idiot, I thought, but I was in the minority. The crowd burst into rip-roaring cheers, and he was crowned the new Mr Nottingham.

I didn't see Ash for six years after university and my enduring memory of him was of that night. In fact, I'd almost entirely forgotten he existed until the spring of 2011, when his profile popped up on Facebook with a friend request. I thought I might as well accept it, and when I viewed his profile, it appeared he'd been leading a rather interesting life.

There were photos of him riding horses in New Zealand, trekking in India and skiing all over the world. I knew he had studied philosophy, but surely he must have a job too? I had recently left the army and was doing a lot of travelling myself. I had not long got back from Africa, where I'd been volunteering for a charity, but now I was out of money and looking for my next project.

Most of my mates from university and the army were now settled down with wives or girlfriends and regular jobs, so I thought about messaging him and seeing what he was up to. He might still be the brash and loud character I remembered from uni, but he seemed to be doing well for himself and had carved out a niche that intrigued me.

A week later, we met for a coffee. He was still larger than life, but had mellowed somewhat and seemed to have his head screwed on. He chatted happily and enthusiastically about how he had worked in every conceivable job going, all around the world, from bartender to cowboy to ski instructor. 'In fact, Lev,' he said, 'are you free next week?'

I was.

'And you were in the army, so you're trained in fighting and all that stuff?'

I tried to explain that it was more nuanced that that, but before I went into any detail, he interrupted, 'Well, I'm just starting a new job running a billionaire's chalet in Verbier, Switzerland. Do you want to come and be the family bodyguard? It'll be good money.'

He didn't need to ask twice. 'Sure, why not?'

So that spring I managed all the security arrangements for a high-net-worth individual and his family. It was hardly what I aspired to, but I needed the cash, and as well as the time it allowed me to ski, it gave me an opportunity to develop the skills I needed to learn to become an explorer and writer. What's more, I got to know Ash properly. Far from the buffoon I had written him off as at university, I found him to be incredibly kind, intelligent and funny. He was polar opposites to me in respect to his outgoing character, but we shared an enthusiasm for travel and debate, and many an evening was engaged in thoughtful, philosophical conversation.

I was glad I had given him a chance and reconnected, and since then Ash and I have travelled together, helped each other out on projects, written articles together and gone on assignments all around the world. Ash was the person who first introduced me to the world of television, and he taught me how to pitch to media outlets. Ultimately, he was the person who inspired me to take a chance on focusing all my energies on writing a book and following my dreams.

Don't write people off simply because they aren't your 'type' on the face of it. Different people bring different things to the table, and it's this variety that gives teams an edge. At university I'd seen Ash as a bit of a showoff, but looking back now on that trip, I became friends with him because he embodies all of the traits of optimism: he does everything with bags of enthusiasm and energy and wears his heart on his sleeve.

The guy who stripped naked in a club and threw a shoe in my face would go on to become one of my most honest critics and one of my greatest friends.

Whānau

The Maori word whānau literally means to 'be born' or 'give birth', and it is often used to refer to the wider family unit. It is compared metaphorically to a spearhead, or a flock of birds flying in V formation. The implication is that, while everyone within the group is free to act as they please, the whole – the thrusting spearhead, the bird formation – is more effective if all its parts are pulling in the same direction.

One of the world's most successful sports teams of all time, the New Zealand rugby union team, aka the All Blacks, consciously adopted this philosophy and added the English short-hand, 'no dickheads'. No one in the All Blacks is bigger than the team, and however talented any given player, egos are not permitted to stand in the way of team cohesion and purpose.

For this reason, after any match or training session, senior players can be found sweeping the changing rooms and clearing away any mess – even after celebrating an important win. Part of their philosophy is: 'No one looks after the All Blacks, the All Blacks look after themselves.'

The most elite Special Forces selection processes around the world have a similar philosophy. It doesn't matter how strong or fast you are – if you think you're bigger than the team, you're out. In the army it doesn't matter which regiment you are in, or what rank you hold, everyone should be prepared and willing to get involved in even the most menial tasks the situation requires – like on expedition. In Afghanistan, the best officers were often to be found filling sandbags alongside the men, or 'stagging on', i.e. standing guard, in the sentry positions. It engendered trust, respect, and cemented the team. If anyone was found slacking, then you'd be considered selfish and ousted.

It takes only a few bad apples to undermine the morale of a team. As the Maori saying goes: He iti wai kōwhao waka e tahuri te waka ('A little water seeping through a small hole may swamp a canoe'). If just one individual doesn't pull his or her weight, then that negativity will inevitably spread through the ranks causing a loss of faith: 'Why should we work hard if they aren't?' Great teams require everyone to unite fully behind the end goal, and work with singular focus towards achieving it.

Attitude and energy

Once we have the self-awareness to confront our own strengths and weaknesses, we can go forward armed with the knowledge of where we are competent and capable, and where we are not. Then, if we don't know what we are doing, we can bring in people who do. This fundamentally underpins the explorer's attitude of mind; no person is an island. There will always be things that you can't do and it will always be better with two. No expeditions are accomplished alone; even if the team is behind the scenes, or is informally made up of friends and family who are fulfilling those roles, it is always a team effort.

When I'm looking to build a team or choose a travelling companion, the first characteristic I consider is attitude. I don't mind if someone doesn't have experience or skills in a particular field, or if they don't know how to read a map. That can all be learned. But if someone is negative or not up for an adventure, then they are not welcome on my team. It's as simple as that. I also want to be around people who are enthusiastic, fun and playful; bringing energy and vitality to what they do. People who are passionate about what they do, without being too earnest or high and mighty.

There is an idea called the energy investment model, which suggests that people can be divided into four different types of behaviour, based on their typical feelings and reactions to events. It is measured by a combination of *attitude*, or how positive someone's mindset towards a certain situation may be, and the level of *effort* or energy someone puts in. The four categories are: spectator, player, deadwood and cynic.

Spectators are characterised by having a positive attitude towards a challenge, but a relatively low energy investment. They are the sort of people who see things through a positive lens and like the way things are; they are unlikely to take the initiative and release large amounts of energy reserves.

Cynics are the opposite, with high energy input but often a more sceptical or negative mental attitude towards the task at hand, often disillusioned about the things they are doing. This means that with their high energy and competence, they can be a distracting or troublesome element. Having said that, leaders do well to listen to cynics, and to take their views on board when their concerns are valid and constructive, because they can be the first people to spot problems in the way things are going. When cynics are given recognition, they can become players; if not, their disillusionment can grow and they can ultimately turn against the group.

Deadwoods have a bad attitude as well as low energy. They feel that things are done *to* them rather than *by* them and in a team they have usually been repeatedly undervalued or unsupported. Of course, these behaviour categories are not set in stone and don't necessarily mean that these members will always remain in this category. Whatever your circumstances, don't accept being deadwood, or a 'victim', as this category is sometimes called. Instead take responsibility for yourself; you are the person with the most power to change your own story. Leaders can also coax better attitude and more energy out of their deadwood teammates by encouraging and supporting them.

Which leads us to the final category, the players, who have a positive attitude and usually invest high energy into challenges and situations. These people tend to take the initiative and are capable of investing the effort required to see things through. It is the responsibility of the enthusiastic, high-energy players to help empower those who are feeling victimised. It's in everyone's power to bring something to the world, and to the people around you, even if it is only a sense of enthusiasm and fun – and players know this. If you can bring humour, warmth and a sense of fun to those around you, that is worth more than any amount of money.

Get to know yourself and which category you tend towards. Players are the sort of optimistic people who choose to get better rather than getting bitter. This stuff is not in the hands of fate; this choice belongs to you.

On all my expeditions, I have to select my local guides wisely. Failure to do so could not only risk the success of the journey, but in many places it could also mean life or death. Travelling across the frontlines of Syria, I chose as my guide a lady called Nada, a straight-talking grandmother, who got me through some of the most terrifying checkpoints I had ever seen. Why? Because as a hijab-wearing woman of a certain age, she appeared non-threatening to the gun-toting militiamen, who were immediately disarmed by her respectful yet strong charisma.

In Central America, I was joined for five months by Alberto – a photographer who had never walked anywhere in his life before. But that didn't matter. I didn't need a guide as such – they are ten-a-penny and I can read a map myself – what I needed was an enthusiastic companion; someone who would be good fun and positive, but could also get us out of tricky situations if we came across them. His natural charm certainly saved my skin on more than one occasion when we bumped into drug runners and gangsters in the Guatemalan badlands.

Team culture

With higher stakes and risk scenarios, team culture on an expedition is apt to be more intense and arguably even more pivotal to success than in everyday life. On expedition, there is total equality and open-mindedness – there is no room for prejudice or intolerance. There are usually all sorts of people involved, each with their own roles: cameraman, director, security advisor, photographer, local fixers, drivers, translators, and so on. Of course it's important that they are focused on their job, but at the same time each one needs to be prepared to get stuck into whatever task is at hand; whether that's helping to put up a tent, washing the dishes or fetching water. Nobody is excluded from these chores unless there's a very good reason.

I've learnt a lot from observing the team culture and team dynamics on expeditions, and teams undergo a process of developing a culture. When it first forms, everyone will busy themselves with defining the scope of the tasks – who will do what, when. Some of this gets discussed openly, and some of it falls into place without anyone having to say it. I've often been in teams where people didn't know each other before, so everyone is busy gathering first impressions and thoughts on each other too.

In these early stages, the group generally get started on tasks on their own, not yet fully comfortable in their relationships with the rest of the team. At this point, all but the most difficult or pig-headed of team members will avoid conflict; they want to be accepted into the group and be well-liked.

This is eventually followed by a period of hostility and conflict; a few frosty words or a row. People start sharing their ideas and their perspectives and this causes disruption or disagreement, and it's a distraction from the work the team is supposed to be doing. Those who naturally avoid conflict tend to find this uncomfortable and will shy away from the collective. Some groups are able to resolve this contention in no time, with good communication, but others will get stuck here, and a few teams never make it past this phase.

If we get the culture wrong, then individuals are forced to waste time on protecting themselves and their own interests from people within their own team. This defence is a waste of precious energy or resources; it causes splinters and factions, and weakens the team. When we feel safe and don't waste time on self-protection, we work together, combining forces to become stronger against the outside dangers, or towards the actual task at hand. Internal competition is perilous; it's hard to act as a team if you're inadvertently competing with your teammates. This is the stage when people start to wonder, *how would I do this differently*, or perhaps, *could I do this better alone?* If you're not careful, your Players can become Cynics.

If teams can get past this, they can get into a phase of cooperation. The team has a shared purpose and shared expectations and after the storms of sharing ideas on how to get things done, a consensus of sorts has been reached. There is a common goal that everyone has in sight. There is a strategy for how the team members might act if conflict rears its head again, so that time won't be wasted on disagreements and distractions. This gives a shared and structured understanding of how things are done, which streamlines and speeds things up enormously.

Only the very best of teams get beyond this into the magical sweet spot of trust and productivity. This is rare, but really rewarding to be part of. There is a synergy between team members, who naturally look out for each other, and the team is performing well with no friction. That's not to say that everyone miraculously gets along and agrees on everything, but group dynamics and interpersonal concerns tend to be aired and resolved quickly and without too much drama.

When this is at its very best, one team member can join or leave, or be absent, and the overall mission – the documentary or the expedition – doesn't suffer. This is a culture that a leader has to foster, but it is also characterised by the fact that team members are not totally over-reliant on the leader for everything.

If you're not all on the same page, your chances of success are severely limited. But if you share a purpose and there is an underlying yearning to serve something larger than the individuals, a team can become greater than the sum of its parts.

Neil Armstrong and NASA

The first step on the moon by a man was also the last of an eight-year odyssey by the largest expedition team in human history. On 20 July 1969, Neil Armstrong had travelled 384,000 kilometres (240,000 miles) in only four days to reach earth's moon and utter those immortal words, 'That's one small step for man; one giant leap for mankind.'

But the Apollo programme that put him and Buzz Aldrin there had employed the skills of 400,000 people for almost ten years. Over 200,000 companies and universities had supplied the equipment and *brainpower. The project cost $24 billion and was the longest, largest and most ambitious expedition ever conceived.*

It is also perhaps the greatest example of how teamwork is fundamental to success. Hundreds of the world's greatest scientific minds came together to solve seemingly impossible problems at breakneck speed to create a rocket and spacecraft that could transport a crew all the way to the moon and back; an astonishing achievement that inspired a whole generation, won the space race and changed the face of human potential forever.

It won't always be plain sailing . . .

Of course, teamwork isn't without its difficulties. Often teams get hamstrung by the personalities within them; egos obstruct and pride gets in the way of making the right decision – stuck in the stormy conflict phase. This is certainly something I've encountered in less formal hierarchies than the army, and on expeditions it can be a real problem, especially when you're tired or frightened, and even if you have the best company.

I'll never forget one particular night in Kashmir. I'd been walking for months along the length of the Himalayas, from Afghanistan into Pakistan and now I had reached the Indian side of the Kashmir line of control. I was with a local horse guide called Mehraj, and my friend Ash Bhardwaj (of Mr Nottingham fame) had flown out to come walking with me for a few weeks in India, the land of his father's ancestors.

We had followed a ridge for most of the day, before descending into a prehistoric forest with enormous ferns and oversized mushrooms. It was the kind of wild, primordial environment where you'd half expect to stumble upon a Stegosaurus at any moment. It had warmed up again and I chatted with Ash about how nice it was to be walking downhill after a debilitating climb the day before. We'd loaded our rucksacks onto the mules and carried only our cameras and water bottles to make things easier.

We had last seen Mehraj a couple of hours before. He was looking after the mules, so told Ash and me to go on ahead. We could wait for him at the top of the ridge above the village of Naranag, where we'd find an open meadow. Apparently it was so obvious we couldn't miss it.

I led the way, following a path that took us deeper into the forest. A thick layer of pine needles covered the track and lightning had felled several trees so that we kept losing the way. It looked as if no one had been here in months, if not years, so there was no way of knowing which of the myriad of trails was the right one. We tried to stay at the same height, but soon enough the path would diverge and then disappear altogether. We would go back and rejoin it and follow another one, but then the same thing would happen.

The afternoon wore on and both of us were getting frustrated. There was no phone signal to contact Mehraj and it was becoming more and more apparent that we were lost.

'For God's sake, Lev,' said Ash in a fit of exasperation. 'Do you know where you're going?'

I showed him my map. It was simple enough to work out where we were and where we needed to be. The only small problem was how to get there. To our left the hill fell away to an almost vertical drop into the valley, where a gushing river chiselled its way through a gorge. To the right the forest was thick and twelve-feet-high thorn bushes blocked our possible route. The village we were aiming for was only five miles away – as the crow flies – but it may as well have been on another planet. I looked at my watch; it was already 5 p.m. and Mehraj would have been waiting for us on the ridge for at least an hour.

'What do you reckon?' I said. 'Up or down?' Both looked equally appalling.

'Let's go down to the river and follow it round,' Ash said. I could tell he was exhausted and in no mood to climb again.

'You heard what Mehraj said. There are no bridges – what if we need to cross?'

I looked down into the valley. It must have been half a mile down, perhaps more. We had two options – a thorny scramble up, or a slippery descent.

I thought for a moment about the options. If I pushed to have my way and I was wrong, then I stood to lose Ash not only as my cameraman, but also potentially as a friend. He had volunteered to be here because of me, and had given up his spare time. Even if he was wrong, then I guess it didn't really matter.

'Okay,' I conceded. 'Let's go down.'

I guessed that it couldn't be too hard, and at least it would be easier than going up again, and so we slipped and slid on our backsides down through the Jurassic-sized ferns. A huge deer darted out of the bushes and seemed to fly into the pine forest below. I wished for a moment that I were an animal – unlike us, they never seemed to tire. We slid further and suddenly an adder slithered from under a rock, its diamond ridges flashing in the undergrowth. His life didn't look as nice as the deer's, and I reconsidered my aspirations for reincarnation.

We reached the bottom of the cliff and emerged onto a narrow beach. It was now almost 6.15 p.m. and my heart sank; we had less than an hour of daylight left. The river was a torrent of freezing glacial melt, crashing through the gorge like an icy bolt of lightning. There was absolutely no way of crossing, and on our side the cliffs up and down the river were sheer faces of crumbling rock. I was exhausted. I hadn't felt like this on the journey so far, but now I felt responsible for leading Ash down the wrong route. Fortunately, he seemed to get a second wind.

'Come on, get a grip,' he said. 'Let's get out of here.'

'But there's no way round,' I pointed out.

'We'll have to climb back up.'

So that's what we tried. Following a gulley, we tried to scramble out of the valley, but the soil was wet and the grass fell away in clumps in our hands. It was too dangerous; one false move and we would fall to certain death in the swirling eddies below. Suddenly a rock came loose above my head and tumbled out of the undergrowth, going straight for Ash's head!

'Watch out!' Somehow I stretched my hands out just in time; it was a miracle I didn't fall myself, but I managed to divert it between my legs and let it go safely past Ash before crashing down into the river. We clung to the cliff where we both shook with horror at the near miss.

'Let's climb back down!'

We had underestimated the power of nature in the mountains and taken them for granted, forgetting that the price of indescribable beauty from afar could mean terrible deadliness up close. It had taken us an hour to slide down half a mile to the river. There was no way we'd get back out before dark. I told Ash and the realisation set in.

'Shit.'

'Exactly.'

We both sat slumped in silence, side by side on a rock, wondering what to do. It was almost dark and the clouds were black. I was so physically done for that I began gasping for air and vomiting, although there was nothing to throw up – we hadn't eaten all day.

'We need shelter,' Ash said.

It was the first rule of survival – especially in the mountains where the weather could change at any moment, and I was grateful that Ash had done some basic army training himself – even if it was only a few weeks as a reservist. He helped me to my feet and we stumbled through the undergrowth, back along the narrow strip of rocks to the way we had come down. A hundred and fifty metres away, almost imperceptible were it not for one angular beam, was the frame of an old poacher's hut.

There is no such thing as a straight line in nature, and when you're in the forest, little things like that can stick out like a sore thumb. The roof had caved in and the whole building was overgrown with wild cannabis and vicious brambles, but it was the nearest thing to shelter we could see. Up close I could see that the only way in was through a gap in the bushes and so we pushed through, getting covered in scratches in the process.

'That was lucky,' said Ash.

'A miracle,' I agreed.

Just as we entered, there was a loud crack and a flash of light. An instant later the rain came down in sheets. 'Sod's law,' Ash looked at me. 'The one time we load the bloody bags onto the mules, you get us lost.'

'Me?' I shot back, annoyed that I'd let him have his way against my better judgement. 'It was your idea to come down the valley.'

'Piss off. You agreed.'

He was right, of course. I should have insisted we carry on climbing, instead of taking the easy road to appease him. After all, I was the leader and bore responsibility. I should have had the integrity to stick to my guns. Now we were stuck in a soaking ravine overnight with no sleeping bags, no food and no way of contacting Mehraj. We were cold, wet and hungry, but at least we had a collapsed roof over our heads and a lighter.

'Fine,' I said. 'Let's just forget it and make a fire.' For the next eight hours, we took it in turns to collect twigs and bits of moss to keep the camp fire alight. I was grateful we did have a lighter, because I didn't fancy my chances of doing a Bear Grylls impression in those conditions. The mere fact that we had a roof above our heads (however leaky) was probably the only reason we didn't die of hypothermia that night. We curled up, shivering around the embers, shifting in disgust as the forest's insects crawled over and around us.

I didn't know if I'd ever been more miserable . . .

Mutiny on the Bounty

Teamwork doesn't come automatically, and leaders can never take the obedience of their teams for granted. The consequences can be dire if, as a result of a leader's actions or their failure to consider the morale of their team, the group ceases to operate as a coherent unit.

Lieutenant William Bligh found this out to his cost during a breadfruit-picking expedition aboard HMS Bounty in 1789. At the voyage's outset two years earlier, Bligh appeared to be a popular leader, raising his men's spirits with regular music and dancing sessions, and reporting that the crew were happy and in obedient order. Upon arrival at Tahiti, where the expedition stayed for five months cultivating breadfruit plants in exchange for gifts presented to the local chiefs, many of the Bounty's crew and officers got involved with Tahitian women. Bligh himself steered clear of this, but permitted his crew to do so for the sake of their morale, on the condition that they continued to go about their duties.

However, over time he became frustrated by their falling standards. He vented his frustration with harsh, often humiliating punishments, especially towards Fletcher Christian, to whom he had previously expressed his approval with a promotion to Acting Lieutenant early in the voyage. Floggings, rare early in the expedition, became routine.

Three crew members deserted, and were flogged themselves once they were recaptured.

When the Bounty left Tahiti, its crew were downbeat after five relaxed, hedonistic months on the island. Bligh himself was becoming paranoid, venting his fears with angry outbursts and disproportionate punishments of his officers, particularly Christian. On 28 April 1789, Christian and a small group of mutineers took over the ship and cast its captain and his supporters to sea in a lifeboat. Bligh and his followers eventually returned to Britain via Indonesia (then the Dutch East Indies). Their expedition had descended into failure thanks to the disorder of the crew and its officers, but Bligh was exonerated from blame.

Christian and his mutineers did not fare well; after capturing the Bounty, they tried to establish their own community on the island of Tubuai, but were forced to abandon this when Christian began to lose control of the team there in the face of skirmishes from local islanders. They split into two factions. One attempted to settle on Tahiti, but were eventually captured by British navy ships set on bringing the mutineers to justice. Christian and his supporters settled on the island of Pitcairn, along with around thirty Tahitian prisoners, mostly women.

Christian's group eventually succumbed to the forces of discord they had unleashed; infighting between the captured Tahitians and the mutineers led to the deaths of all but one of the latter. Christian was killed while tending his fields, but his son, Thursday October Christian, survived. The population that they founded still lives there to this day.

. . . but it's better as a team

Despite sharing the hut with some hairy spiders and millipedes the size of eels, Ash and I made it through till dawn. At last the rain abated and a chorus of birds heralded a new day; one in which we were alive. Weak with hunger, we climbed back out of the valley, following our trampled course from the day before, hoping to retrace our steps and find our guide, Mehraj.

Fortunately for us, our Kashmiri guide had the initiative to send out a search party. After just an hour, we heard shouts from the woods up above. I looked up to discover an old shepherd wielding an enormous axe and waiting for us on top of a boulder. I had never been so relieved to see anyone in my life.

I learned a lot of lessons from that experience. It was a mistake that cost us a miserable night, and yet if Ash hadn't been with me and I had got lost alone, things could have been far worse. I have always enjoyed travelling on my own. It gives you time to think, and some of my most useful journeys have been the ones that I completed alone. It forces you to interact with people in the place you are visiting, as you don't settle into the comfort of just chatting to your travel companions.

But there are perils, as evidenced by the sobering tale of Chris McCandless, who spent two years wandering, hitchhiking and hiking alone in North America. It was made famous by Jon Krakauer's retelling of Chris's diary entries in his book *Into the Wild*, which later became a film. Young Chris headed into the Alaskan wilderness without a map and in the hope of living in a peaceful solitude. But after four months alone in the forest, he finally succumbed to starvation. His diary entry sums up the dying man's final thoughts: 'Happiness is only real when shared.'

I was glad that I hadn't gone through that night in the forest alone. Sure it was a screw-up, but now Ash and I look back at it and laugh. So long as people have faith in the journey and believe in the team, sometimes the biggest mistakes can yield the best results. We remain the best of friends.

Choose wisely

I was once given some wise advice; that in order to achieve great things, you need to surround yourself with great people. Whatever it is you are setting out to accomplish, think carefully about the people that will be best placed to help you get there. Be honest with yourself about your own shortcomings, and think about the people you know who might be able to plug those gaps. In turn, work out what you can contribute that those people can't; it is a team game, which means it's not all about what you can get from other people, but how you can maximise your contribution to the whole, as well.

Above all, give your team time to establish its culture, and *listen* to the advice of the people you've brought on board. Far too many leaders, especially in the business world, ignore the words of Apple CEO Steve Jobs: 'It doesn't make sense to hire smart people and tell them what to do: we hire smart people so they can tell us what to do.'

Motivational speaker Jim Rohn has said that you are the average of the five people with whom you spend the most time. If that is the case, then choose your friends wisely and make sure you don't let your ego and pride dictate those that constitute your circle.

Teamwork is at the heart of great achievement, and fundamental to success. I've learned from my experiences in the army, in business and in the world of expeditions that getting the right people on board with your vision is the best chance you have of succeeding in any task, project or mission. Teamwork is about removing your pride, and working in harmony with others towards a common goal. Good teamwork is underpinned by the selfless actions of the individuals within the team, and this can be inspired through having a solid vision and a plan of how to enact it. It's about building mutual respect and trust and it is brought to fruition through making it fun and enjoyable. There is great joy to be found in being part of a team, particularly when on the road.

Remember that there's no such thing as a one-man expedition. Solo holidays are one thing, but if you are trekking to the North Pole, even if it's on your own, you will still need back-up in the form of a team at home tracking your progress, or providing emergency support. On all my big expeditions I've had to seek assistance in some form or another, whether that has come in the shape of resupply drops in the jungle, or getting local guides, fixers and translators to help navigate the complexities of travelling in difficult places. It becomes even more necessary to have a team of good people around you when you have a specific mission in mind, like making a film or documentary.

As part of a team, we can achieve great things – effective teams, pulling in the same direction, are more than the sum of their parts. Once we're able to operate as team players, we are able to take the next step towards our own individual goals, and we can have one of the most fulfilling and rewarding experiences that life has to offer – that of being a leader.

Mary Seacole

7

Lead From the Front

I used to say of him that his presence on the field made the difference of forty thousand men – Arthur Wellesley, 1st Duke of Wellington, describing Napoleon

The secret of leadership

Helicopter blades slashed through the hot sky, the roar of engines deafening as the pilot twisted and turned on our final approach. Crammed into the back with my men, I tried to keep my balance by tensing my legs, and focusing my gaze on the machine gun in the helicopter's door. My kit was heavy on my shoulders, and sweat poured into my eyes from under the rim of my helmet. My palms were wet as they gripped the rifle. I pulled it closer to my chest, and prepared myself for what was to come.

The airman behind the machine gun turned to me. His face was covered by a scarf and the dark visor of his helmet, but I knew what it meant when he held up his index finger.

I turned, and shouted to my men. 'One minute!'

One minute to landing, and whatever was waiting there for us.

Looking into the eyes of my young soldiers, I saw steely resolve – the hardness of the paratrooper flying into battle. I was closest to the rear door, which was now winching fully open. There was no doubt in my mind that I had to be the first one down the ramp. I had waited years for this moment, and now was my time. I'd practised it over and over again, on grass fields, in mock-up structures and on the real things during exercises all over the world, but this was different. For the first time in my life, I was flying into war.

Today, there would be an enemy waiting for us.

For the last eighteen months, I'd been training with the thirty men of my platoon for this moment. For eighteen months before that, I'd been schooled in the art of war and leadership at Sandhurst, and at the Infantry Training School in Brecon. I had three years of the best military training in the world to my name, but it would mean nothing if I wasn't the first down the ramp.

There was no way of knowing exactly what waited for us. It was the spring of 2008, and the fighting season was commencing in the deserts and valleys of southern Afghanistan. Since we'd arrived a month before, there had been the occasional patrol, and we'd seen mortars exploding in the far distance, but as a platoon we hadn't yet been 'in contact'. We hadn't seen with our own eyes the Taliban that we'd been sent to fight.

We were heading straight into the enemy heartland. A place called the Arghandab valley. It lay outside of Kandahar city, the birthplace of the Taliban, in the notorious *Dasht-i-Margo*, which is the Pashtun term for the 'Desert of Death'.

Our mission was a simple one – to kill or capture a Taliban bomb maker called Haji Mohammed, infamous for making the improvised bombs that were responsible for the deaths of dozens of British and allied soldiers.

We'd spent a week poring over maps and aerial photographs of Nalgham village, in which he was thought to live. My platoon – No. 8 Platoon, on attachment to A Company of the Third Battalion the Parachute Regiment – was given the honour of landing first, right in Haji Mohammed's back garden, and it was our task to surround his house.

It was a dangerous mission, but as the seconds counted down, and I looked out of the rear door of the helicopter towards the dusty plains below, I felt a tremendous sensation of both trepidation and sheer excitement. It was a lot of responsibility to shoulder, but I wanted to carry it.

The helicopter got closer to the ground, whipping the Afghan dust into the air until it was a thick cloud around us.

'Ten seconds!' shouted the airman, holding tightly to a rope. The door was now fully open. The ramp hit the floor with a clunk, and a shudder went through the airframe as the wheels touched down – everyone wobbled but kept their feet. They knew how important it was to stay upright, and get off the heli quickly. The Taliban had spies everywhere; scouts that reported the movement of helicopters. Every second that we delayed deplaning was a second for the enemy to train their weapons and kill those on the ramp.

Being on that ramp was my rightful place. There are times when an officer needs to step back to better control his men, but this wasn't one of them. I would be the first off this aircraft, and into whatever waited.

'Go, go, go!' the airman shouted, myself and my men picking up the call so that it rippled through the aircraft. And then we were running, gritting our teeth as though that would stop the bullets that may await us.

My feet touched down on the Afghan soil and I breathed in the dirt as I ran through the cloud that the helicopter's blade stirred up all around us. My rifle was up at eye level and I looked over my sights, ready to snap shoot at anyone that posed a threat to my men.

My soldiers followed me. I felt like a giant, leading them from a metal beast and into the jaws of death . . .

Except that when the helicopter lifted away, and the dust settled, we were quite alone.

Nobody home.

Shit.

I put some of my men into defensive positions and took others with me to search the house. This was still a dangerous time, as the Taliban were not above booby-trapping their own homes. Other than a few sacks of opium – the drug of choice in those parts – we found nothing.

I walked back outside and was shrugging off the disappointment of another quiet mission when a whip-like sound echoed across the fields, followed immediately by a crack. It wasn't the noise you hear in the films, it was altogether more visceral and unnerving.

'Incoming!' shouted my platoon sergeant.

I shouted at the men to jump into a nearby ditch and return fire. They looked at me, unbelieving. It was like being on an exercise on Salisbury plain, except the noises were not the dull simulations we'd become accustomed to. These were real and violent. Somebody was shooting with the evident intention of killing us.

Three hundred metres away, I saw the enemy moving in a line of trees. I took aim, and fired.

It was the start of a very long day.

Helen Sharman

When I was thirteen, I went on a school trip to the Birmingham confer-ence centre. Despite the dull sounding venue, this particular trip was very exciting, not just because we were allowed to wear normal clothes and potentially meet girls from another school, but because we were going to hear a real-life astronaut speak to us about their experiences.

In 1991, Helen Sharman became the first Briton in space. Her story is remarkable. Just a couple of years before, in November 1989, this young woman was studying for a doctorate in food chemistry when she heard an advert on the radio. It was calling for volunteers to go into space – no prior experience needed – as part of a Soviet mission called Project Juno. The idea was to improve relations between Britain and her Cold War adversary.

Helen applied and had to undergo masses of physical and mental tests, from being whizzed about in a centrifuge to aptitude tests and problem-solving exams. Out of 13,000 applicants, she was one of two who made it to Moscow for training. She had to learn Russian, of course, so she could communicate with her fellow adventurers, and after eighteen months she was picked to go on the mission.

So at just twenty-seven years old, the young Yorkshire woman became the first British person to launch into space for an eight-day flight. Like lots of youngsters, I was fascinated by space, curious more about the mundane intricacies of daily life on a space mission than any great ques-tions about the profundity of space travel. And that's what she described. She told us about where they sleep, how they eat and what it is like to float around in zero gravity. She explained the challenges she faced – from space sickness to muscle wastage – and told us what it was like to look out of the window and gaze down on our planet from afar.

In my mind, astronauts and cosmonauts were superheroes. Yet here was this very normal woman, who seemed like one of my teachers, or my mum, talking about how she had been into space. I was transfixed by how ordinary she was – not in a rude way, but her stories were so relatable. It was a revelation to me; previously explorers had belonged to far-off distant lands or story books, but here was a nice lady from Sheffield right in front of me, who was also an astronaut. That was when I realised that there are no super-humans, everyone is capable of phenomenal and earth-shattering things; anyone is capable of being a pioneering leader – in any field.

In the questions afterwards, one little girl asked her if she had been afraid before the launch. She was very matter of fact: she said, there is no need to be afraid if you know what you're doing. She said the only astronauts she knew of who were scared or nervous before their space flight were the ones who didn't fully understand the procedure, or know what was going on, as they were usually passengers for the launch phase and part of the mission later on.

She knew she was well-trained and more than adequately equipped to handle it and so she wasn't afraid. That is a powerful lesson, in that with practice and hard work comes skill, and in that confidence is born the basis for strong leadership.

It's not a popularity contest

Leadership is one of those words that gets bandied around quite a lot, but what defines a good leader? They come in all shapes and sizes, and there isn't a simple formula to explain what makes a good leader.

Politicians, activists, generals, coaches and business tycoons have all written great books about the types and ingredients of leadership, and there isn't enough space to do it justice here. Leadership will mean different things to different people, and my understanding of what leadership means is heavily informed by my experiences in the army, as well as my experiences founding a business and leading teams in hostile environments.

What I can tell you for certain is that leadership is a crucial part of the explorer's mindset. Being trailblazers and boundary pushers, explorers are inherently leaders, and are looked up to. There is also the more practical matter of testing expeditions that require exemplary and proficient leadership in order to succeed and to get everyone home in one piece. For me, leadership is about setting an example to those I have the honour of leading. Ultimately, it's about inspiring people.

Of course, leadership is not limited to the battlefield. The same lessons apply in any situation. Whether you're in a sports team or a busy office, setting an example to your team is essential to engendering trust and respect; for without it, you're not a leader, you're a dictator.

When travelling, I've found myself in various circumstances that have required me to take a stance and demonstrate some leadership. Sometimes it was requested, and other times it was down to my own judgement. Often, it is hard to know when one must stand up to the plate, especially in a situation where there is no set hierarchy, and it's even more difficult when there is already a leader in place, but they aren't doing their job.

When you are in a team as part of an expedition, there will inevitably be times when you have to set an example, even if you aren't the designated boss. I've been on a number of expeditions where it might simply be a case of volunteering to tidy up the camp; to go and fetch water, or put up the tents. These are clearly the building blocks of what it takes to be a good team player, but being a good team player is also an attribute of a good leader. Others will look to your example and emulate it, which is ultimately the underlying aim. Let's not beat around the bush – leadership is about getting other people to do what you want them to do, and sometimes those things might not be pleasant.

In the business world, this might mean delegating a particular task, or asking an employee in your growing team to fire a colleague. In the army it's arguably more serious, in that people might be putting their lives on the line. Good leadership is not about likeability. There is a big difference between good leadership and popularity, though there is often some crossover, but it's important to understand that they are not the same thing. In fact, I'd say that a fixation on being liked can end up being to the detriment of respect; people can see through a desperate bid for approval and it can eclipse the real priorities. It is the difference between making decisions based on what will go down well with the group, compared to what is the right thing to do.

To take a less extreme example than the military, entrepreneurs are often said to have a reputation for being disagreeable; the sort of people who aren't reliant on the approval of their community and contemporaries. More likely to flout convention, they have to get used to saying or doing the unusual or non-conformist thing, and this takes a certain type of person; someone who is a natural pioneer, and does not seek approval from others telling them they are right.

Of course, different leaders have distinct ways of coming at this. Some people have a leadership style that is enormously charismatic and heavily dependent upon personality. The obvious advantage of such a leader is that those being led are imbued with energy and enthusiasm because, quite simply, they like that person. These leaders are motivational and followers want to listen to them, be around them, and importantly, follow them. But there's a major disadvantage to this reliance upon likeability; if the liked leader leaves, the group no longer has a figurehead. A group, a mission or an entire organisation could be pinned on the appeal and popularity of this one leader.

Additionally, any leader will sooner or later have to make an unpopular decision, or one where it's impossible to keep everybody happy. Leaders who rely on likeability for their authority will quickly crumble in these circumstances, as with our earlier story of Captain Bligh. Research shows that there is a gender bias here, too; success and likeability are positively correlated for men, while they are negatively correlated for women, which comes with its own set of unique complications and challenges.

One of the most liked entrepreneurs of our time is Elon Musk. Again and again he's made decisions that leave others aghast, but engender him to others. Smoking weed on the Joe Rogan podcast may have dropped Tesla's share prices the next day, but millions of people decided that they liked him based on that interview and action, and Musk knows how to turn that following into leadership capital, and business.

Musk is well known for asking the opinions of his staff at all levels. Many other leaders seek out popularity with their followers by adopting a far more devolved or participatory style of leadership, wherein the leader engages their followers in the decisions they are making. By getting people involved in the room, the leader is including followers in the broader mission, but it can be challenging and inefficient when trying to reach a consensus.

Others go so far as to be totally hands-off in their approach to leading, providing little or no guidance or supervision. This can make leaders popular – not meddling too much in the lives of their followers, or micro-managing and directing their actions – but what if those who are being led have little experience, or are seeking a figurehead? This approach would be untenable in the military, where leadership tends to be more hierarchical and commanding, in order to ensure efficiency where safety is tantamount.

This distinction is important. Very few entrepreneurs are likely to lead their employees through actual life-or-death scenarios, and certainly not in the split-second decision-making context that combat entails, so civilian leadership plays by a different rulebook from military leadership.

In the military there's no time for dissent; soldiers have to get used to following orders unquestioningly, especially under pressure. In business, it pays to have everyone challenging the consensus a little more, thinking outside the box and being prepared to voice opinions that go against what the CEO might be thinking – otherwise, board-rooms simply turn into groups of yes-men. Even in a military context, giving subordinates a degree of autonomy is beneficial. However, good leaders foster this level of dissent and autonomy for the creative ideation it facilitates, not for the popularity they think it gives them.

Live to lead another day

In October 2009, I led an expedition to climb Mera Peak in the Nepali Himalayas. At 6,476 metres (21,247 feet) it is one of the highest non-technical summits in the world, and we were a bunch of novice climbers. Only two of the team had ever worn crampons before, and they did not include me. That said, we were all paratroopers, imbued not only with some of the hardest training in the British military, but we had also recently returned from Afghanistan and were high on life, looking for the next adventure.

After a couple of days in Kathmandu, where we prepared our gear and bought last-minute supplies, we managed to get a flight to the tiny, high-altitude airstrip at Lukla, the gateway to Everest and the mighty Khumbu region. From there it was going to be a two-week round hike up the wildly remote Hinku valley, where we would camp amid the enormous boulders of the glacial moraine. Apart from me, only my senior NCO Geordie Taylor had been to Nepal before, and for the other seven soldiers, the sight of local children drinking directly from a yak's udder was bewildering, to say the least.

As the days progressed, the walking got harder with each metre of altitude gained, and even the fittest of the team found themselves out of breath. It was my job, as the sole commissioned officer and the expedition leader, to show them how it was done. Even when I was exhausted, I tried to flit between the front and the back, checking to make sure everyone else was okay, often when it was the last thing I wanted to do.

After more than ten days of hard trekking we finally reached the snow line, where temperatures plummeted to minus 27°C, and with gale-force winds making the setting-up of camp a dreadful prospect. At least three of the others and I were suffering from severe head-aches, dehydration and very cold feet, as we tried unsuccessfully to sleep at 5,000 metres.

On the summit day we were greeted with clear conditions, but two of the local Sherpas decided to remain with the tents in camp rather than ascend to the top. I led the way with Geordie, and our mountain guide, Jason. We plodded through the thick snow, tied together in case a crevasse opened up. It was punishing, slow going, and every step was hard-earned with short, gasping breaths. The summit came into view, a few hundred metres ahead, but there was no time to celebrate.

As we stopped for a short break, one of the soldiers approached me complaining that he couldn't feel his feet any more. His boots had been on the tight side, and as a result his toes were right up against the edge of the boot. As we'd been trained. I ordered him to remove his boots and socks, and I placed his bare toe inside my jacket right up in my armpit, where it would be able to warm up to body temperature naturally.

As we waited, I saw another soldier shifting about. He told me his feet too were numb, and so after a while I offered the same service to him, but before long my feet were also getting so cold they were excruciating. We needed to move, so once we'd taken a rest, we pushed on another hundred metres to a crag. But then, just before the final push to the summit, disaster struck. A howling wind flared up the valley, followed immediately by a white blanket of cloud. I knew instantly that even if it was short-lived, it would add time on to our exposure, which might mean the difference between some-one keeping or losing their toes to frostbite.

We had trekked for days, and planned for months. No one wanted to give up

'Boss!' one the soldiers shouted. 'We can make it, it's not far to go. Two hundred metres more!'

I looked into the murky abyss of frozen air, and then back at the faces of the two shivering soldiers. I was torn. From the looks around me, it was clear the overwhelming urge of the group was to continue. These were all paratroopers, men proud of their own hardiness and expecting it above all from me, their leader. But I knew that if I decided to continue, I was putting all of our safety in danger, and the fault would be mine alone for any consequences.

'We turn back,' I shouted. 'Now!'

I hauled the youngest of the crew to his feet and pointed down the slope, back to the camp where the Sherpas waited for us. The most headstrong of the soldiers muttered and shook his head. I didn't say anything or blame him; I was as disappointed as he was. We all knew that our chance of conquering the mountain was gone, at least for this expedition. Months of anticipation, and weeks of physical training had gone to waste.

As we got back to the tents and packed up the camp, all were silent. It wasn't until hours later when we descended beyond the snow line, and to a sheltered gully where the sun had warmed up the rocks, that one of the senior soldiers approached me.

'You made the right call there, boss. Otherwise we might have died. Or at the very least, we'd all be missing a few toes.'

As we carried on down the trail, the disappointment that we all felt at not achieving our aim seemed to dissipate, and we were glad to have come down unscathed, having lived to fight another day.

I had not wanted to take the decision to turn back. It felt like failure, and I worried about the impact it would have on my soldiers' estimation of me. But as I discovered afterwards, even those who had wanted to continue at the time had a lot of respect for the decision in hindsight.

Moreover, that defeat instilled a wonderful attitude in those eight soldiers. After the expected banter and complaining that all soldiers are good at, every single one of them said that the trip was one of the highlights of their military career. What's more, many of them later became serious climbers and mountaineers, and went back again to summit Mera Peak.

If I had made the popular decision that day, rather than the right one, it could have been a very different story.

John Le Marchant

The British Army is rightly seen across the world as a bastion of exemplary leadership, but it wasn't always the case. In fact, if it weren't for the determination and vision of one particular military commander, it might never have been so.

John Le Marchant was a mid-ranking officer during Britain's calamitous Low Countries campaign of 1793–95. During the campaign, he was struck by the lack of leadership and equipment in the British Army, particularly its cavalry, which contributed to a significant defeat for Britain and its allies. He admired the horsemanship of the allied Austrians, but was aggrieved to hear one of their officers scoffing that British swordsmanship was 'most entertaining', but reminiscent of 'someone chopping wood'.

Le Marchant was clearly stung by these comments, and on returning home he set himself about lifting the standards of the British Army. To my mind, this is a defining trait of leaders; they don't settle for inadequacy, but are always looking for opportunities to improve things, and taking steps themselves to do so. He designed a new cavalry sabre – a heavy sword – with the help of cutler Henry Osborn, and published in 1796 a celebrated manual on cavalry swordsmanship; he then toured Britain and Ireland instructing cavalrymen on how to use it. However, his most enduring mark in the field of military leadership was yet to be made.

In 1801, despite initial resistance from parliament on the grounds of cost, Le Marchant established the British Army's first training school for officers that would lead foot soldiers – known as the infantry – and cavalry. This new establishment was titled the Royal Military College at Sandhurst, and 205 years later, I would stand on its parade square.

Le Marchant served as Sandhurst's governor for its first nine years, and was personally responsible for training many of the new and existing officers who would go on to serve under Wellington in Britain's much more successful part in the Peninsular War, which led to Napoleon's defeat in Spain and Portugal, and his imprisonment on the small island of Elba.

In 1947, Sandhurst merged with the Royal Military Academy, Woolwich, to form the Royal Military Academy, Sandhurst, whose mission is to be 'the national centre of excellence for leadership'.

Serve to Lead

Serve to Lead is the motto of the Royal Military Academy Sandhurst. The motto isn't mindless waffle – it underpins the mentality of the British officer class that has led soldiers on operations around the world for generations. Sandhurst teaches that service is fundamental to leadership and only by making sacrifice and leading by example can someone earn the trust, and right to command, of those under that leadership.

While some leaders are born, the academy also teaches that leadership can be learned by those willing to undergo rigorous training and preparation for the responsibility of command. It is the place that I learned how to be a leader too, and where I discovered both the joys and hardships of that duty, which I subsequently fulfilled on missions around the world and especially the battlefields of Afghanistan.

Before you can ask others to do your bidding, you must first be willing to do it yourself. My old brigade commander, and later Chief of the General Staff, Lt. Gen. Sir Mark Carleton-Smith, once gave all the junior officers in the Paras a fireside chat. He told us that the very best leaders should be able to do everything that their subordinates can do, and better.

That's obviously a big ask, and it isn't as clearly applicable in every walk of life. The CEO of a big tech company doesn't necessarily need to know how to code, for example, as long as he understands the business; and the conductor of an orchestra doesn't have to be able to play the cello. However, if they could, then the coder and the cellist would certainly be very impressed, and consequently would probably work that little bit harder.

Of course, teamwork and leadership are intertwined. You don't have to be a leader to be part of a team, but you do have to be part of a team to be a leader. One of the hardest parts of being a leader can be striking the right balance between the two roles. I have found, though, that the line can be walked so long as there is trust and loyalty.

Loyalty goes in many directions. In the army and in big organisations, there is loyalty to your superiors – those that give you orders; then there is loyalty to your peers – your friends, colleagues and acquaintances; and most importantly perhaps, there is loyalty to those under your command or in your team – because if you're not loyal to them, they will be the cause of your downfall. If people trust you, however, and they believe in your allegiance, then they will follow you to the ends of the earth.

Mission command

In the army there is a doctrine called mission command. It is all about the art of delegation and giving people responsibility for their own actions. It's based on the principle that when you manage a team, you should tell people *what* to do, but not *how* to do it. So long as they are qualified to do a job, you have to put your trust in their training and that they are capable of making their own decisions, which shows loyalty and engenders a real sense of ownership. That is how to build a team.

Once, on operations in Zabul province in southern Afghanistan, my platoon was given the mission to search a village for a known terrorist recruiter. For the most part, this involved being invited in by friendly women for tea, scanning their kitchens and gardens with metal detectors, and then being cursed and told the house belonged to an absent uncle as we dug up, and removed, caches of weapons.

After walking around all day, we were low on water and needed to get back to the helicopter pick-up point, which was five miles away across the desert. As we were leaving the village, I noticed a group of men huddled around under the shade of a mulberry tree. They looked shifty and stared at us as we walked past. One of them wore a white turban and had kohl around his eyes, looking like a Taliban leader. I got the translator to say hello and ask their names, which he duly noted.

I reported the names by radio to our intelligence cell and they said they were all clear, so we left the men and trekked back across the desert to where my boss was waiting. By now the whole platoon was exhausted and thirsty. We had not had time to eat all day and the temperature was over 50°C. One of the soldiers was beginning to wobble and I suspected it might be heat exhaustion.

Then, to make matters worse, my boss came over. 'Lev, you know that group of men in the village?'

'Yes?'

'Well, the one in the white turban, the Int cell now want him in for questioning.'

'What the hell? I reported his name on the radio and they said he was clear.'

'I know. They screwed up.'

'So what now?'

'Well, we've got two hours before your heli comes. Go and get him.'

I shook my head. I had two hours and it was 10 miles – a 16-kilometre round trip. In the Paras we have a well-known physical stamina test known as the Ten Miler, which is a punishing speed march while carrying kit. They can be hard work at home in the rain, but out in the desert and low on water, I knew that it might be deadly.

I looked at my men and could tell that half of them weren't up to it. They were exhausted, but orders were orders. I figured that I could do the job with ten or twelve men, so half the men could stay behind. I knew that in order to do so, I had to get them to buy into the vision and feel ownership, so instead of barking out an order, I gathered my section commanders around. I told them the situation and asked them what they thought were the best options.

A lesson I learned early on is that even if you already know the answer, ask the question – it makes people feel valued and part of the team and decision-making process. It doesn't matter who gets the credit for the decision, and when you take your own ego out of the equation, it's amazing what happens.

'Sir, I have an idea,' said the youngest corporal. 'Why don't we leave half the men behind, and take some of their water. I'm sure we can do the job with ten or twelve men.'

'That's a great idea,' I said, patting him on the back. He grinned from ear to ear, and I made him the point man.

'I need twelve volunteers to come with me.' I said. 'The rest of you can stay here.'

The men looked at each other. They knew it would be one of the hardest tabs of their life and that it would be dangerous, because the Taliban now knew our strength and would have time to prepare an attack as they saw us walking back across the desert.

Private Foster, one of the new soldiers, put his hand up. 'I'll come, sir. I need the exercise,' he joked. Sylvester was next, then another and another. Because we had a strong team, bonded with trust, I had no shortage of volunteers. Even some of the men whom I knew stood no chance of making it started to put up their hands, because they felt ashamed, but I already had enough. We redistributed the water, gritted our teeth and marched back across the desert to the village where we found the man in the white turban, arrested him, and marched all the way back again, being chased by an angry mob of Afghans.

It was a hard slog, but one of the most determined team efforts I'd seen in my career. To top it off, we later found out that the man in the white turban was at the top of the regional most-wanted list. That victory, and a shared hardship, cemented the bonds of the platoon even stronger – there was nothing that my men didn't think they could accomplish.

Mary Seacole

Born in 1805, Mary Seacole was a British-Jamaican nurse, healer and businesswoman, who became best known for her contribution to the Crimean War, where she established the 'British Hotel' behind the lines – a place Seacole described as 'a mess-table and comfortable quarters for sick and convalescent officers'.

Seacole did not have formal medical training, but she applied the healing skills that she had used in Jamaica, and was arguably the first nurse practitioner. She had originally been refused permission to travel to the Crimea as part of the war effort, but she was not one to take no for an answer, and made her own way there to offer her services to the injured soldiers. Mary became incredibly popular with the troops, because of her relentless, selfless commitment. So much so, that in 1857 crowds of 80,000 attended her fundraising galas in London.

Mary Seacole left a lasting impression on the medical world, and Britain's National Health Service has a leadership programme named in her honour. It teaches students valuable lessons in self-awareness and emotional intelligence; how to find a leadership style that suits your personality; how to wield authority and motivate a team; and how to transform emotion into an asset – all things that Mary learned on the battlefield and yet are applicable anywhere.

Mary Seacole is remembered to this day as one of the most formidable and yet caring leaders in the medical world.

Raising the stakes

Courage is at the heart of good leadership, and I don't mean solely in terms of taking measured risks. Whilst courage can come in many forms, it is mainly about accepting the truth, and being honest with yourself and others. The most frightening and yet rewarding place to be is when you are completely and brutally honest with yourself. It is both the hardest and best gift you can ever give to yourself – and others.

In being ruthless in your assessment of your own character, of your flaws and failings, and also your strengths, you are far better placed to accept the truth of the circumstances around you, and therefore how best to react to them. In doing so, you give yourself a greater chance of success in whatever it is you hope to achieve the first time around, rather than making mistakes through denial. What's more, when a leader is able to remove his or her ego from the decision-making process, it will engender trust among the team.

Whatever context you are leading in, you have a singular responsibility to those you are leading, and this may mean that speaking the truth is more important than receiving the praise. To be a good leader, you need to have the moral courage to tell people not only what they want to hear. For a military leader, your subordinates' lives are at stake, but for business leaders, the same is true of their livelihoods, and as far as their families are concerned, this verges on the existential as well. CEOs have a responsibility to their employees to make good decisions as opposed to popular ones; if they make bad decisions and the company goes bust, they are taking food off their followers' tables.

Above all else, remember that to be respected, leaders have to be *respectful*. Many leaders forget that although the objective is not to be liked, or likeable for its own sake, this isn't a green light for leaders to treat those below and around them without dignity. In fact, this is one of the quickest ways to turn your players into cynics. Remember always to understand, value and appreciate the distinct abilities and perspectives that each of your team members is bringing to the table, and respect the courage it takes for them to voice an opinion that goes against your own.

You need three things to bring people with you and compel others to understand and buy into your vision.

The first is all about you, the leader, as an individual. Be the person you pretend to be, and live up to the expectations you have of others. Of course, there is always a certain amount of 'branding' – what you decide to say and how you package yourself will define how others view you. It is important to consider what you will choose to share or amplify, and what will fall away because it doesn't serve the vision. Be honest, but you don't always need to let your team know everything; enigma can sometimes be a very powerful tool. What is most important is that however you choose to display your attributes, you can act on what you say. Style is nothing, unless it is underpinned by substance.

The second is about utilising your emotional intelligence to appeal to the collective and its shared purpose. This is about knowing your audience, and a good leader will always tailor and adjust what they say according to who is in the room. In some instances, it is ideal to pitch the tone as if you were chatting to a friend – it's amiable and warm and can build trust. In others, such as in the army, it would be inappropriate, and a more formal tone might be better suited. However you do it, this is about identifying and bringing out shared values and communicating them explicitly in your vision.

Finally, leadership is about getting things done. It is about pushing through your will and achieving the goal. A leader who can talk the talk but not walk the walk is not a leader, and so often the critical mindset is one of understanding urgency. If there is a deadline – meet it. Great leaders compel people to participate in their vision right away, so that the job gets done, come what may. It might not be the perfect solution, in fact it rarely is, but as we were always told in the army – an 80 per cent solution done on time, is better than a 100 per cent solution late.

It can be lonely as a leader. Sometimes the path might not be clear and you will question your own choices. But, if you truly want to be a pioneer, leading the way, exploring the world's hidden places, then it's a necessary path to take. How then, do leaders stay on course and navigate the perils ahead? The answer, as we'll find in the next chapter, relies on a well-calibrated moral compass.

Captain Lawrence Oates

8

Keep an Eye on the Compass

I am just going outside and may be some time
 – Lawrence Oates, Antarctica, 1912

Watch your step

In Afghanistan we often came under attack from the Taliban. They would snipe at our patrols and launch mortars at our bases, but I think I speak for most soldiers when I say that the scariest aspect of being on operations was the prospect of hidden roadside bombs and landmines. In a firefight there is an element of control, and the ability to hit back with one's own weapon, artillery and air support, but for the professional modern foot soldier, the biggest terror is the unknown – any footstep could be your last.

In addition to the homemade bombs created by the likes of Mr Mohammed and his amateur engineer friends, the whole country was littered with landmines – a legacy of thirty years of war, where many were left behind by the Russians in the 1980s and never cleared up. Often, the minefields were never properly marked, and to make matters worse, over the decades the mines washed away in the rains and snow and moved from their original locations, so it's often impossible to predict where they might be.

Every patrol was a game of Russian roulette, so to speak, and we never knew if we would come back alive. One day, as we were setting up a temporary encampment on the outskirts of a lush valley, I got the call from my commanding officer that a vehicle had been blown up a couple of miles away and I was to take my platoon on a rescue mission.

Our armoured cars had special tracks, so that they could go over any kind of terrain. We drove them across the desert and up to the top of an escarpment, where the location of the explosion had been reported. As we got closer, I looked through the front windscreen and could see the remains of a Land Rover blown to smithereens. I was expecting the worst, as there should have been four soldiers inside. I called to the team medic and told him to be ready to treat any surviving casualties.

We stopped short by fifty metres in case there were any other bombs nearby, and when I got out of my vehicle, to my absolute surprise and joy, I saw that all the men were fine. When the mine exploded, it had sent a shockwave through the car, sending everyone in it flying out onto the ground – they were all alive, and seemingly unharmed. I asked them if they were okay, and apart from being a bit dizzy and shaken, they nodded and walked over to where I was standing. There was space in one of our three vehicles, so I put the men in the back of the car at the front, telling the lead vehicle that we should reverse back out of the danger area, because we might be in a minefield.

It seemed like the sensible option, so that's what we did, making sure that we stayed exactly in our own tracks until we were well clear of the flat plateau. At that point, my three vehicles needed to turn around so that we didn't get stuck in a bottleneck between two large cliffs, so the front vehicle did a three-point turn and pushed around the other two so that it was now facing forwards, again in the lead. We all did the same and drove off down a track back towards the encampment, and I was now in vehicle number two, where the commander should always be. Then just as we left the scraggy boulder field, there was an enormous explosion, and I felt the shockwaves hit my own car.

After a second of deafness and ringing in my ears, I looked through my windscreen to see a massive cloud of dust. As it cleared, I realised what had happened – the point vehicle had driven over another mine – and to make matters worse, it was the same four soldiers who had been in the first explosion. I shouted down the radio telling everyone to stop exactly where they were whilst I considered the situation.

I could see bodies lying in the dirt next to the car, which had the front end ripped off it. Now I knew I had to show some real leadership and there was only one thing for it. The metal detectors were in the car at the back and it would take them a good ten minutes to clear the path between the cars, let alone get to the front, where we might have heavy casualties. I got out of my car, pulled out my bayonet and began to crawl forwards, stabbing the sand in front of me to check for the mines. Luckily, the car in front wasn't too far and I was able to cover the distance quickly. I reached the car and to my relief, found that yet again, by some miracle everyone was alive, and what's more, uninjured.

One of the soldiers stood up, dusted himself off and looked at me stabbing around in the dirt.

'What are you doing sir?' he laughed. 'You won't find any Taliban down there!'

It was a lucky day, made even luckier as I was filling out the daily report back at camp, when I found out that my team and the remaining vehicle had been tasked with 'denying' the half-blown up cars; that meant going back to the site and blowing them up properly with explosives, so the Taliban couldn't make use of them in the future. It was a simple task and I was told that I should remain behind with the other platoon commanders, so that we could receive our orders for the next day. As the team were driving back, they too hit a landmine and the empty commander's seat, where I would normally have been sitting, was completely destroyed by the explosion.

You can't be a leader unless you're willing to put your team ahead of yourself when circumstances call for it. This requires moral courage and integrity, the building blocks not only of great leadership, but of a fulfilled, purposeful life.

Headhunters

In the isolated foothills of the north-eastern Indian border with Myanmar live a tribe who, up until relatively recently, had a very different moral compass and sense of integrity than our own. The Konyak warrior tribe is one of many Naga tribes in the region. However, what makes them different is that up until the 1960s, the Konyaks performed an important cultural practice – headhunting.

Any dispute over territory was settled with warfare and the Konyaks were famed and feared for taking the heads of their enemies. Taking a person's head in battle was understood as taking their power. The heads would be brought back to the villages as trophies, and presented on doorways or walls, or carried in specially made baskets. The warriors would also be tattooed after battle to mark them with their heroic deeds.

Since the 1960s, the Indian government has made headhunting illegal, but there are some people alive in the villages today who have the proud tattoos from battle. They still carry the long knives, passed down through generations, that were used for the violent act of removing the head from the body. Even now at festivals and rituals, members of the Konyaks will bring out their collected heads of the past to remind others of their strength and bravery.

In the West, we associate the act of beheading with criminal punishment in past years and, more lately, the work of terrorists. However, for some people, headhunting was until very recently a show of a man's worth and virtue. One man's barbarism is another's norm.

Moral courage and integrity

Moral courage is about doing the right thing even if it's the hard thing. Integrity is about doing the right thing even when nobody's looking. That being said, what those morals are can be difficult to define. This makes perfect sense, since these personal principles are so subjective, entirely shaped by our nature and our nurture. Each of us has a unique understanding of right and wrong, which has come about from every one of our experiences and interactions; an endless number of factors determining our current belief system. To a certain extent, our moral compass is a personalised construct; not that it feels that way when you see or hear something repulsive, or terrifying, that strikes you as evil and so obviously inhuman or 'wrong'.

On occasion, I have felt this when travelling, where different communities have their distinct cultures and entirely separate belief systems. If we could travel in time, no doubt we would be horrified by much of what supposedly upstanding individuals got up to. In my travels, I have visited people with very divergent customs to the ones with which I was brought up, and found their ideas and actions sometimes distasteful.

Different cultures have their own moral codes. Some US states consider the death penalty acceptable; Islam allows a man to have four wives, whereas in Bhutan a woman may marry two brothers if she chooses; across Africa, female circumcision is widely practised. In parts of China and Vietnam people eat dog meat; in France and Russia, they eat horses.

In Britain, any of the above would be considered not only illegal, but utterly immoral. Then again, some of the things that Brits do would be considered very disagreeable elsewhere. Getting boozed up on a Friday afternoon, whilst entirely appropriate on a bank holiday in London, would go down rather badly in Bangladesh or Saudi Arabia, as would our fondness for bacon sandwiches and pork scratchings.

I've had lunch with fighters from Hezbollah, a designated terrorist group; I've met plenty of men from cultures that embrace polygamy, and I've even met an Aghori monk who thought that cannibalism was perfectly acceptable. I didn't agree with their standards and would not condone their actions.

I do however believe in dialogue and communication, and I've found that by engaging with people whose views are diametrically different to your own, it can benefit both parties. You aren't necessarily going to change someone's mind over an issue, but by living according to your own values and by setting a good example, you might broaden their horizons a little. What's more, you will inevitably broaden your own. I never went on a journey with an agenda, or an intention to change or save the world. I have always tried to be an objective observer and refrain from judgement. By meeting those whose moral compass is oriented differently from your own, you simply develop a more sophisticated understanding of your personal moral code, and it may even swing your needle in one direction or another.

Good and evil

In the West, we've become used to thinking of Good and Evil as two opposing qualities, and the concepts are so second nature to us that sometimes we forget they need defining. It can seem self-evident what 'good' means and what is 'evil', until we remember how easily politicians can manipulate whole groups of fundamentally 'good' people into culture wars over issues that blur the boundaries. On top of this, that very notion of 'goodness' being opposed to 'badness' is the product of about 3,000 years of Old Testament influence on Western religious and moral thought.

The Genesis story of Satan tempting Eve with an apple, which God had forbidden her and Adam from eating, lays the foundations of a moral code in which Good (obedience, restraint, faith and humility) is diametrically opposed to and competes with Evil (diso-bedience, greed, deception, dishonesty and hubris).

Not all cultures see things the same way. Chinese Confucianism, for example, is more concerned with the proper rules of behaviour for educated or higher-class people. It's less about what we'd call Good and Evil, and more about Right and Wrong. Buddhists see Good and Evil as opposed to one another, but believe that both must be overcome and recognised as non-existent concepts, in order to achieve Sunyata or 'emptiness'.

There's a big difference, though, between morals and rules, and this is an important distinction to me. Our morals are often governed closely by rules and ethics, such as the justice system in the country we grew up in, and the laws to which we adhere, but they're not the same. The best leaders and people with the highest integrity stick unwaver-ingly to the former, their morals, but are often the first to break the accepted 'rules'. Indeed, someone once said that rules are for the guid-ance of the wise and the obedience of fools, and are often guilty of being outdated, throwbacks from moral codes of yesteryear.

However, I believe there are a few universal morals that are impor-tant to adopt in life, and particularly when exploring and travelling. One of these is respect for others. Many of the worst aspects of travel in the modern world stem from a lack of respect; people treat the countries of other people very differently from how they would treat their own home. It's pretty basic, but the tiny amount of forethought involved in bringing your own canteen and some water purifying tablets will stop you having to litter parts of the world without clean drinking water with endless plastic bottles.

Another is generosity. We need to remember that it's a privilege to be able to travel, or even explore outside the boundaries of our own upbringing, and that many of the world's population don't get this opportunity. We need to be thankful for it, and venture out with the openness of spirit that enables us to give back.

Whatever your moral values, though, sticking to them in the face of adversity is what really counts.

No shortcuts

In 2012, I led an expedition to walk across Madagascar. This was perhaps the most challenging of all the short-form expeditions that Tom and I ran with Secret Compass: a three-week trek from the east coast to the west coast of Madagascar, through some of the thickest rainforest I'd ever encountered.

The team consisted of eleven paying clients, me, the expedition assistant, Alastair, and a doctor, my friend Kate. I'd also employed the services of a local guide and a few porters to help carry our food and tents. It was an uncharted route over the highest mountains on the island. I'd chosen it deliberately because it looked like it would be tough, and also because it would enable us to see the remaining wilderness jungle on an island where 90 per cent of the original rainforest has been cut down.

After a crossing from the African mainland, endless red and umber hills poked out above the low-lying mist as we flew into Antananarivo. Snaking brown rivers weaved in between treeless valleys and only the rising smoke of village fires and dusty roads broke the mesmerising undulation. Flying over the stripped hillsides of the central highlands certainly gave us the impression that we may have been too late to see the original rainforest – and I admit that I was a little disappointed before the journey had even begun.

The plan was simple – to lead the first group on a walk across the northern part of the island from east to west, entirely on foot. It was a distance of almost 400 kilometres in the seemingly impossible timeframe of three weeks – a daunting task. All this in the company of strangers, many of whom had never travelled anywhere so remote.

'This is my first time anywhere near a jungle,' said Xaviar, with a nervous smile. The rest of the group looked at the 27-year-old French Canadian with sympathy.

Days passed by as we trekked west. As the settlements thinned out, the jungle grew closer. A high wall of black vegetation loomed ever nearer in the shape of an impenetrable mountain range. On the seventh day of the trek, the rainforest finally surrounded us. It had receded a full 20 kilometres since 1962 – the last time the region was properly mapped before Google Earth. Despite its foreboding appearance, it came as a relief to be amongst the trees.

This is what we had come to see, the Madagascar of our imagination. There were no people here. The tracks disappeared and off we went, hacking and traipsing our way through primary vegetation in one of the last remaining pieces of true forest on the island. For the next week, the group was entirely encased in green vines, hairy bamboo and wet ferns. Elusive lemurs stared curiously down from their canopy hideaways, as we ploughed on through marshes and rivers. It was one of the toughest treks I'd ever done, and that included most of the ones I'd accomplished in the army.

As we neared the western coast, villages sprung out of the cultivated valleys and the jungle at once became a hazy memory. All eyes were set on our goal – to reach the port of Ankify and make history. The final day turned into a true test of integrity, not to mention mental and physical endurance: there was still 93 kilometres to push, and we were a day behind schedule, having taken longer than expected to fight through the jungle. Delaying wasn't an option, as people had flights to catch back home and jobs that awaited them.

I studied the map: 93 kilometres is 58 miles. It was a long way, and everyone was already exhausted. I looked at the team. I knew there was a trailhead not too far away, where there would probably be a village with cars and motorbikes; from there we could scoot to the beach and enjoy a night out before flying home the next day. I wasn't about to make the suggestion first, though. Luckily, Simon, one of the other team members did.

'Look, why don't we just hitch a lift for the final day, we've almost done it anyway, and we don't have anything left to prove do we?' Almost everyone else nodded, glad that he had offered them a way out. He glanced at me as I remained silent.

Then I saw Xaviar shaking his head. 'We've come this far,' he said, 'and I paid to walk across Madagascar.'

He was right, of course, and I knew I had to demonstrate some leadership, despite the fact that every bone in my body wanted to jump in a car and get to the beach as soon as possible. I didn't have anything to prove. I'd done my time in the army and slogged through endless months of hardcore expeditions already. This was supposed to be fun!

And yet, something inside me knew the right thing to do.

'Those who want to carry on walking, come with me. Those that want to get to the road, go with Simon.' I looked around, my legs hoping that Xaviar would see the light and choose to follow the crowd.

'I'll walk,' said Kate, the doctor.

'Me too,' said Alastair.

The decision was made. I put on my most enthusiastic face. 'We're walking then.'

And that's what we did. As the rest of the group walked to the road and got a taxi to the beach, the four of us, and a couple of the local guides, set off at 2 a.m. to hike for 19 hours on what was one of the most punishing days of my life. By the end of it, when we finally reached the coast at almost 10 p.m., we were a bedraggled bunch with blisters atop of blisters. Alastair could barely walk and Xaviar was cramped up in agony.

But we'd made it.

Looking back, I didn't have to do that final day's walk, and I'm sure if I had suggested to Xaviar that we should all take a car to be on the safe side, then he would have followed. But deep down I knew what was right, and it was to attempt to finish the job, rather than taking the easy way out. What is more, everyone in that group looked up to me in some way, or was relying on me in some other way.

When you are a leader, you are letting more than yourself down if you don't adhere to your morals and act with integrity. Only by pushing myself, and the team, to their absolute limit were we able to achieve something that everyone else thought was impossible, and it is a lesson that I have often reminded myself of later in my career. It was a principle that got me through much bigger and more ambitious expeditions in due course.

This is why integrity is the cornerstone of leadership. When you lead, you are setting a tone and a culture. Not just for yourself, but for those around you who have been compelled to follow you in some way. This might be tiny in the grand scheme of things, or it could be far more substantial or influential; even the founder of a business transmits aspects of their own moral compass to a whole organisation.

In setting the tone, you have also taken up a mantle of sorts, a responsibility to those whom you are leading. It is your duty to do the best by yourself and by those people.

Captain 'Black Sam' Bellamy

Our moral compasses don't always equate to the law – though, by and large, they should! Someone who demonstrates this rather nicely is the famed pirate and explorer Captain 'Black Sam' Bellamy, who was born in Devon in the seventeenth century. There is no denying that pirates were the ultimate rulebreakers of the day, rebels who made a living from sailing the high seas and stealing from other ships.

But Black Sam was a slightly different sort of pirate, in that he had a pretty sturdy moral compass. He refused to wear the long, powdered wigs that were so fashionable and associated with the wealthy merchants he so despised, sticking instead to his natural long black hair, which is what earned him his name. He described the Navy and the establishment in his diaries: 'They vilify us, the scoundrels do, when there is only this difference: they rob the poor under the cover of law, forsooth, and we plunder the rich under the protection of our own courage.'

Sam wrote often of his conscience and was a charismatic captain, known for the mercy that he showed his victims; he and his men would often take ships without harming the crew, which was unheard of at the time. In fact, if they took a ship on the high seas and discovered that it did not suit their purpose, they would return it with crew and cargo intact. Captain Black Sam ran his ship as a republic, allowing his team to vote in important decisions, and he was known by his men as the Robin Hood of the seas.

He died very young in a storm, at only twenty-eight, on a ship with an almost 150-strong crew and an estimated four tonnes of loot, which now sits at the bottom of the ocean. Captain Black Sam Bellamy knew above all else that if you want to achieve greatness, you've got to stop asking permission.

Morals or rules

Sam Bellamy was a rule-breaker, but he had a sound moral compass. I'm not advocating breaking the rules for the sake of it here, despite the words of His Holiness the Dalai Lama: 'Know the rules well, so you can break them effectively.' But the lesson we can take from the story of Black Sam is that moral courage and integrity allow you to achieve more than you could without them.

As leaders, as explorers and as people, we are faced almost daily with challenges to which there is no clear answer. This is going to happen even more so in the future, as the work that humans, rather than machines, do will increasingly be that relating to other people – work that requires us not to be cold and calculating, but warm, considerate and creative. Without moral courage and integrity, this kind of ambiguity – dealing with people and their in-built unpredictability, or pursuing creative goals where something hasn't been done before – is impossible to navigate without some kind of guiding framework.

This is why we use the metaphor of a compass when we talk about morals. They guide us, helping us to navigate through previously unexplored territory. Without morals, we are rudderless. With them, even if it's impossible to see the way ahead, we have a set of values and actions to which we can commit, and trust that they will take us where we want to go.

And so we come onto integrity – doing the right thing when no one is looking, and when no one will know about it. My *moral values* determine who I am and what I stand for, and they give me a benchmark to understand right and wrong; *integrity* is whether I follow or practise those moral values. When a leader acts with integrity, they are adhering to their own moral code consistently, upholding the same principles in all situations and not wavering. They also follow it honestly and wholly, doing what they say they're going to do.

We are acting with integrity when our belief systems and our actions match up. It is often hard to ascertain the results of decisions, and we never know what might have been if different choices had been made. So, in the absence of a crystal ball to predict the future, one needs to establish one's own set of rules to live by, and then at least when it comes to decision-making you have a framework to go by – that's integrity.

You can choose which values and standards to abide by, as they are fundamentally a personal choice. Having integrity is about intentionally following your morals and choosing your own constraints. It will stop you from being pushed around by whatever forces happen to be strongest in your life and allow you to guide yourself. These should become a part of you and your personality; they are the sole things that ought to define who you are. Really, this should be the biggest lesson in life.

Doing the right thing can sometimes be hard. But you don't have to be faced with life or death situations to demonstrate moral courage. I've had my own integrity tested plenty of times on the road, and it is often in the moments when you think you can get away with things – those smaller, everyday incidents – that your integrity will be most tested. Of course, for others, one's integrity can become a matter of life or death.

I May Be Some Time

Some people seem born to a life of setbacks. Lawrence Oates was one of them. He lost his father when he was sixteen, and had to drop out of Eton College after only two years due to an illness. He was able to make enough of a recovery to join the British Army, and served as a young officer in the Boer War. When aged twenty-one he was shot in the thigh, and for the rest of his life his wounded leg would be an inch shorter than the other.

Even at a young age, Oates had an unshakable purpose. During one battle he was told twice to surrender, but replied simply, 'We came to fight, not surrender.' His actions that day brought him into consideration for the Victoria Cross, Britain's highest award for 'conspicuous bravery, or some daring or pre-eminent act of valour or self-sacrifice, or extreme devotion to duty in the presence of the enemy'.

Only four years after his near-death experiences at war, Oates applied to join the Terra Nova expedition to the South Pole. Referring back to our team-member types, we might describe Oates as the 'Cynic' on that journey; he was at times scathing about the expedition's leader, Robert Falcon Scott, especially his relative lack of knowledge about travelling with horses. However, he attributed much of his harshness to the challenges of the conditions they encountered, and was kept loyal to the expedition by a sense of duty, writing: 'Myself, I . . . would chuck the whole thing if it were not that we are a British expedition.'

The expedition was successful in reaching its goal, but discovered to their dismay the remains of Norwegian Roald Amundsen's camp, the rival expedition having beaten the British to the South Pole by 35 days. Dejected, the party turned around and started retracing their 79-day journey. They were quickly beset by extreme weather conditions, disease and food shortages, and lost a team member – Edgar Evans – in a bad fall.

Oates was beginning to suffer from seriously frostbitten feet, and his old war wound had reopened due to scurvy. His rapidly deteriorating condition and pace of walking was putting the rest of the party increasingly behind schedule, which meant that they were falling short of the pace required to reach their predetermined food stops. The whole party faced the risk of starvation, because of their soldier's accelerating demise.

On 15 March 1912 – almost two months since they had turned around at the Norwegian camp at the South Pole – Oates told his team to leave him behind in his sleeping bag. He knew that he was putting them all in danger, but his companions refused to abandon him. Robert Scott's diary then records Oates famously taking matters into his own hands. During a severe blizzard, he announced, 'I am just going outside and may be some time.' Scott's diary records: 'We knew that poor Oates was walking to his death, but though we tried to dissuade him, we knew it was the act of a brave man and an English gentleman.'

Oates couldn't have known whether his companions would make it back without him, and tragically, they did not. All Oates knew for sure at the time was that, with him hindering their progress, none of them stood any chance of survival. His moral compass instilled in him a determination to give his teammates the best chance of survival he possibly could.

There is a memorial placed near where he is presumed to have died, his body having never been discovered. The inscription reads:

Hereabouts died a very gallant gentleman, Captain L.E.G. Oates, of the Inniskilling Dragoons. In March 1912, returning from the Pole, he walked willingly to his death in a blizzard to try and save his comrades, beset by hardships.

Staying true

Moral courage and integrity will help you in any walk of life. As a leader, they will inspire the respect of your team, and give you a framework to help you make difficult decisions. As an explorer, they will help you to make a positive impact wherever you go. But even in day-to-day life, outside the contexts of leadership or exploration, a few mindful practices that cultivate moral courage and integrity will have a really beneficial impact on your life.

For example, you might find your job frustrating and unfulfilling. It can help to take a step back and ask why this is, with particular reference to your moral code. Think about the values and principles that matter to you in life: do you want to help poorer people that you don't know, or to provide as much as you can for your family? Both of these are perfectly valid moral stances, but will probably take your career in very different directions, so it's important to be honest with yourself about which is more important to you.

You should also identify any moral red lines you're unwilling to cross. For example, you might decide that providing for your family is the most important thing to you, but not at the expense of exploiting less fortunate people elsewhere. This can help you identify why you find your work unfulfilling: are you being asked by your manager to do things that you think are morally dubious? Does the business you're in contradict your moral values?

If so, the answer is not to jump ship, at least not straight away: first, write down any beneficial changes that you or your business could make, and suggest these to your manager. This is the stuff that leaders are made from; they effect positive change from the inside. It can be challenging, but however unlikely it seems, your superiors will respect the fact that you've taken a moral stand. If they don't, this is a pretty good sign that their moral frameworks and yours are irreconcilable, in which case you can think about what you want to do instead.

Teams and individuals always perform best when they believe they are working towards a higher purpose. So don't compromise on your values; you will never achieve your potential unless you believe 100 per cent in the value of the task you have set yourself. Use the power of integrity to bring yourself towards your goal. If you want to become a better runner, commit to a certain distance of practice per week, and hold yourself accountable for it. Don't be swayed by the fact that no one will know if you skip one run; if you really want to accomplish something, what matters is how you behave when no one is looking.

Moral courage means the conviction and will to identify and stand up for what we believe in: integrity is the commitment to doing this day in, day out, even when no one is watching. This isn't easy: it takes motivation, tenacity and resilience. With these qualities, you can weather any storm.

Lt. Col. Sir David Stirling

9

Carry On and Grit Your Teeth

*Do not judge me by my success, judge me by how many times
I fell down and got back up again* – Nelson Mandela

You've got to be in it to win it

In the winter of 2006, the unit that I was about to join, 3 Para, had recently returned from their first tour of Afghanistan. It was a deployment that will go down in history as the one that started a decade of violent conflict, and was some of the heaviest fighting involving British soldiers in a generation.

Along with the other new officers, I took over a platoon of battle-hardened veterans, and suddenly being their boss was a daunting prospect. To give us young officers something useful to do, and to get the men reintegrated back into 'normal life' in the UK, our commanding officer, Stuart Tootal, ordered all the junior commanders to pick a sport or activity and organise training sessions. I recall John Martin was good at rugby, so he gathered up all the blokes who enjoyed that and assembled a team. Matt Clamp helped to set up the football team. Tom Bodkin, who was a proficient skier, led the ski team, and so on.

As for myself, I was an enthusiastic boxer, but the Paras already had a boxing team that was regarded as the best in the army, so that wasn't an option. I decided to try my hand at something entirely new and told the CO that I would organise a Parachute Regiment kayak and canoe team.

Colonel Tootal chuckled. 'You do know that Para's hate water, don't you? Good luck.'

So, I started to beg, borrow and steal all the necessary equipment – boats, life preservers, paddles, and all the rest. Luckily one of the gym sergeants volunteered to help me out in the training, as he had done it before. Getting volunteers was a rather trickier process. Most of the soldiers had already signed up to one of the other team sports and, as Tootal had advised, most Paras had no particular interest in getting cold and wet – especially given that it was the winter months, and our only available water was the Chelmsford canal system.

After some persuasion, I managed to get some less than willing soldiers to join up, but that was only because Tootal had ordered that every man must join one of the teams.

As an afterthought, the amused Colonel decreed that rather than doing the sport for a jolly, each of the teams must enter some sort of competition and – being the Paras – do a bloody good job of trying to win. Suddenly, what was until then a bit of a lark became serious. None of the young platoon commanders wanted to lose. We had all just arrived and needed to make an impression, but given the fact that the men had returned from a very hard tour of duty in Afghanistan, they weren't exactly enthusiastic about training in what was supposed to be their down time.

Anyway, the rugby lads all signed up for an international match against the Australian Army. The football team entered the tri-service annual competition, and Tom Bodkin took the guys off to ski in the brutal downhill army championships. Canoeing, as ever, proved to be a difficult proposition, not least because most of us had never really paddled before. Tootal told us that we had until the Easter holidays to complete training, enter whichever competition and come back having upheld the Regiment's long-standing reputation for being the best at everything.

I suddenly regretted choosing kayaking. For the next three months, I took our band of merry men into the icy, shopping-trolley filled canals of Essex to learn how to paddle, Eskimo roll and avoid drowning in the murky depths. Once, by way of sheer accident, one of the soldiers flipped his boat and emerged having caught a fish.

Needless to say, despite my best efforts, none of us was particularly good at the sport, let alone ready for a competition.

I searched high and low for something suitable to enter. The Devizes–Westminster race would have been the obvious choice – a hard slog along the length of the River Thames, attracting the best kayakers in the world to try and finish all 220 miles in under 24 hrs. But thankfully that wasn't until May, outside of the commanding officer's timeframe, as real soldiering was to begin in earnest after the spring. The only competition I could find advertised in the army adventurous training schedule was the Army Canoe Polo Championships, to be held at Aldershot swimming pool on Easter weekend.

It was make or break. So with that, I lobbed a football into the canal and we began to practise, much to the amusement of the Chelmsford homeless population, one of whom decided to take a piss off the town centre bridge and onto the heads of my heroes in training. That didn't end well, and neither did our training. It would be an exaggeration to say that the men were motivated. Not only had they been coerced into a sport they weren't interested in, after a long and gruelling tour of Afghanistan, what made matters worse was that the competition was slap bang in the middle of Easter leave.

I'm sure every single one of the 'volunteers' would rather have been at home with their families, or out boozing with their mates, but when I asked if anyone wanted to drop out, there was silence. Thankfully the sense of camaraderie that we had forged out of the ridiculous situation was strong enough to engender a competitive spirit.

So, on Easter Sunday we travelled by minibus to Aldershot, the spiritual home of the Parachute Regiment, and waited at the swimming pool for the competition to start. At this stage we stood no chance of winning anything, given that many regiments took this sport seriously and trained year round in actual swimming pools, but we readied ourselves to put our best paddles forward and at least try to retain a little dignity in defeat.

A sergeant of the Physical Training Corps arrived at the poolside with a whistle around his neck, looking bored. He seemed surprised to see us.

'I thought everyone had given up,' he said. 'All the other teams have called in sick. Lazy bastards didn't want to lose their leave.'

I was suddenly taken aback. 'Well, we are here and we are ready to play.'

The sergeant shrugged his shoulders. 'Okay,' he said, pointing at the pool, 'get in the boats.'

So we all got into the little sports kayaks and lined up at our side of the pool, waiting for another team to arrive.

The PT sergeant looked at his watch and blew his whistle.

'Well done, Paras. You can get out now.'

Perplexed, I watched as the sergeant came up to us one by one. He shook our hands and gave us all a gold medal – we had become the army canoe polo champions by default.

Not that anybody else had to know that . . .

We returned back to the barracks triumphant. I was the only one of the young officers to have led their team to victory: the commanding officer was over the moon with me, and my 'volunteers' went home early with their gold medals to brag to their girlfriends.

Never give up. It's not over till it's over.

Tenzing Norgay

To Sherpas, the world's highest mountain is known as Chomolungma or 'Holy Mother'. Westerners called it Everest, and in 1935 a young Sherpa called Tenzing Norgay was selected as part of Eric Shipton's reconnaissance expedition. His inclusion came from a bit of luck, as two of the expedition's first choices failed their medical tests, and Norgay's winning smile allegedly caught Shipton's eye.

Following this first venture onto the mountain, the young Sherpa would take part in five more unsuccessful attempts on the summit. Almost twenty years after his first climb on Everest, Norgay was selected to lead the twenty Sherpas who, in addition to nearly 400 porters, would accompany Colonel John Hunt's expedition in 1953.

By this time, Norgay had earned a formidable reputation for himself. George Band, the youngest climber on the expedition, described him as 'the best-known Sherpa climber and a mountaineer of world standing'. Norgay's previous assault on the mountain, a 1952 expedition with a Swiss team, had set a new altitude record for attempts on the peak. Edmund Hillary said of him in 1975: 'His success in the past had given him great physical confidence . . . Tenzing had substantially greater personal ambition than any Sherpa I had met.'

As well as being a determined and diligent climber, Norgay was a smart, outside-the-box thinker. In 1947, he managed to cross mid-partition India by train, unchallenged and without a ticket, by wearing one of his old employer's British Army uniforms. His ability to think on his feet served his mountaineering career brilliantly. Early in the 1953 expedition, he saved Hillary's life by attaching a rope to an ice pick after the latter fell into a ravine. This moment led Hillary to insist on him as his climbing partner for any future summit attempt.

On 29 May 1953 – which Norgay later came to celebrate as his birthday, knowing only that he'd been born in late May – he bore the weight of his fellow climber, Hillary, as the latter reached the summit of the world's highest mountain, the Sherpa soon following him up. As he described it in his autobiography Mountain Man, 'A little below the summit, Hillary and I stopped . . . I was not thinking of "first" and "second". I did not say to myself, "There is a golden apple up there. I will push Hillary aside and run for it." We went on slowly, steadily. And then we were there.'

Norgay described his life as a mountaineer as 'a long road . . . From a mountain coolie, a bearer of loads, to a wearer of a coat with rows of medals who is carried about in planes and worries about income tax.' His self-deprecation notwithstanding, his inspiring journey towards one of the world's most inaccessible points is a testament to the values of hard work, tenacity and diligence. He stuck to his task, always kept his end goal in sight, and was rewarded for it.

Grit

Albert Einstein once said, 'It's not that I'm so smart, it's just that I stay with problems longer.' Ask almost any successful person – in almost any field – what the secret to their accomplishment is, and you will likely receive the same answer again and again. Grit – determination, tenacity and diligence – is the deciding factor between winning and losing, more so than natural talent.

As Professor Angela Duckworth, who studies grit and self-control at the University of Pennsylvania, puts it, success 'entails diligently working towards challenges and being able to maintain effort and interest over long periods of time (years) despite setbacks and stagnation in progress.'

It impacts our happiness, too. A Harvard University study in 1940 asked 130 20-year-old men to run for up to five minutes on steep, fast treadmills. On average they only lasted four minutes, and some managed only a minute and a half. The participants were then interviewed every two years for the next six decades. Those who showed more determination on the treadmill at twenty years old were statistically more likely to have led fulfilling, happy and successful lives since then.

Duckworth and her colleagues developed a Grit Scale, which people can use to measure their own powers of determination. She studied the achievement of various groups of people in competitive circumstances, including the grades of Ivy League university students, the drop-out rates of West Point Military Academy cadets and the scores of National Spelling Bee candidates. Across all groups, grit and determination were better indicators of success than natural ability or IQ. Central to this, in Duckworth's opinion, is that grit equips people with the drive to 'stay the course' in the face of challenges and setbacks.

Resilience

Resilience is our ability to face crises or adversity, and remain calm and healthy in the process. There are countless factors that contribute towards resilience, most of them rooted in our childhood circumstances, but one of the most intriguing factors is positive emotions. Humour, optimism, goal-orientation and even forgiveness have all been shown, in scientific studies, to have a correlation with our resilience to stressful situations.

The field of cognitive behavioural therapy (CBT) aims at increasing people's resilience to stressful situations by building these kinds of positive emotions and thought patterns, replacing negative thoughts that reduce our resilience, such as *I can't do X*, with positive thoughts like *I can do X*. This small difference in our underlying outlook about the world can have big consequences for how ready we are to face adversity. Resilience is also built through coping skills, such as meditation, exercise and socialising. Keeping our minds and our bodies healthy and happy has a huge impact on our ability to overcome difficult challenges.

Socialisation is a massive part of this, particularly in military scenarios. A 2009 study by the US military into the reduction of stress factors for soldiers found that unit cohesion and morale were the best predictors of a unit's combat resiliency. Units that fostered effective peer support and cohesion were less likely to suffer stress-related breakdowns under pressure, and better able to respond adaptively to stressful circumstances.

Keeping standards

This ability to respond to stress calmly, effectively and adaptably is vital in the military, especially for officers and elite units like the Paras and Special Forces. Any decision made – even in training – can lead to the death of others, and so it is vital to 'keep your head when all about you are losing theirs', as Kipling put it.

To join the Parachute Regiment, candidates must undergo a tough selection process called Pre-Parachute Selection run by Pegasus Company, which is usually referred to as P Coy or P Company, based at the Infantry Training Centre in Catterick, North Yorkshire. After 21 weeks of training, candidates are put through eight tests designed to test their resilience and determination, including a 20 mile endurance march laden with a 16 kg pack and a rifle in under 4 hours and 10 minutes; an intimidating assault course 17 metres above the ground called the Trainasium; and 60 seconds of milling.

Milling, arguably the flagship event of parachute selection, is a boxing match between candidates of similar size and strength in which determination and aggression are rewarded, and dodging and blocking lose points. In short, it amounts to being punched hard in the head for a minute straight while trying to hit the other guy as much and as hard as possible. If either combatant sheds blood or is knocked to the floor, the clock is paused, blood wiped off and the bout resumes. It is forbidden to aim at any part of the opponent's body besides the head, and the winner is the most aggressive candidate.

Milling is designed specifically to replicate 'the conditions of stress and personal qualities required in a combat situation' and test the 'determination and raw fighting spirit' of P Company candidates. It is this raw spirit of controlled aggression that sets the Paras apart from the rest of the army.

Nothing worth having comes easily

Nothing worth having comes easy, and we will need to work hard to achieve the things we want most out of life. Whether this is learning a new skill or advancement in our professional careers, we need to be prepared to stay the course and keep working towards our goals, no matter which obstacles appear along the way.

When I was on my way into the Parachute Regiment, I bussed up to Catterick for my Pre-Parachute Selection. At that point, I had no idea what the next few weeks would look like for me; all I knew was that they would be the most stressful, demanding and challenging of my life so far. But I had a goal in mind; I was determined to earn that maroon beret at the end, the recognition of my achievement and my initiation into one of the world's elite fighting regiments.

That level of drive is absolutely crucial. We can't give our all to anything if we're not motivated by the end result. And if we're not giving our all, we've no chance of sticking things through when the going gets tough.

Over the first few weeks of Selection, it was pretty clear that a lot of people didn't have what it took. Some couldn't handle the rigours of training and fell behind; others decided that enough was enough and dropped out of their own accord midway through. I had my own moment when I thought my time on Selection was up. I came down with a shin injury that set me back behind the rest of my cohort for a few days. Unable to join in the rigorous drills and knowing that when I was back in action things would be exponentially tougher for the time I'd missed, I wondered whether this was all worth it.

I started thinking about what my life would be like if I failed Selection. Probably not all that bad; I could probably go and reapply to one of the army's other, more sedate regiments. Then what? Have a fairly average military career, perhaps go to some interesting places and do some interesting things. But nothing grabbed me about it. Even more so when I considered a life and career outside the army; having just passed out of Sandhurst, I couldn't for a moment contemplate the prospect of a civilian life.

All I wanted was that maroon beret. It kept appearing in front of me, every time I closed my eyes. In a way, it would have taken some motivation to present myself to one of the officers running Selection

CARRY ON AND GRIT YOUR TEETH

and tell them I wanted to drop out; that I couldn't do it. But I had no desire to do anything besides recover, get back to training and persevere through the rest of selection.

So that's what I did. I pushed myself harder than I'd previously thought possible, quickly making up the ground I'd lost on the rest of the cadets. I remember my fear of heights being pushed to the background when I tackled the Trainasium. My sixty seconds of milling passed in such a blur that I can't now remember them at all, besides an overwhelming need not to hurt the person I was paired with, but to prove to him and everyone looking on that I wasn't one for backing down, whatever the circumstances.

Perhaps I'm fortunate that I was born with this amount of grit and determination. It's not always a positive; I've been told many times that I am as stubborn as a mule. But up there in cold, wet, windy Catterick, these inner reserves of drive, vision, grit and resilience pulled me through and I got the beret I'd been dreaming about. It is hard to think how different my life would have been without it.

Who Dares Wins

Today, Special Forces are something that everyone is aware of. You hear about them on the news, and see veterans of their ranks on the television, but back in the Second World War the idea of 'SF' was the unpopular brain child of a man called David Stirling. Serving in North Africa, he was frustrated at the pace of action against the Germans and Italians, and the senior command's failure to embrace the concept of commando soldiers.

Stirling was of the view that his superiors were giving up on the idea too easily, and he conceived the SAS, a secret unit that could launch covert surprise attacks in the desert, by night. Stirling finally persuaded his seniors to buy into his vision by sneaking straight into HQ, and marching into the office of the person who was highest up the command chain – an action that was utterly forbidden in the rigid hierarchies of the day. Convinced, the command allowed Stirling to assemble his first team, and now the SAS is famed for its work in special operations and counter-terrorism.

Back then, Stirling set about handpicking a team that could create maximum damage for minimum cost, and these men came to be known as 'the originals'. Stirling targeted those who were frequently getting into trouble in the army for being naturally rebellious or free-spirited, approaching them personally and asking them to join his quirky band of brothers. True pirates, all of them had a thirst for action. Some of them naturally very calm and considered, and others explosively violent. It is thought that more than a couple were likely to be psychopaths.

Stirling managed to recruit a rugby player of enormous stature with a drinking problem, a very smart Oxford-educated Scotsman who invented a bomb they could attach to planes, a priest who came on missions but refused to carry a gun and a former diplomat with fluent Italian. They were all exceptionally brave, tough as nails and willing – come hell or high water – to commit with dogged determination to the mission. Without these traits, you weren't eligible.

Life expectancy was extremely low for the early recruits and the team's first mission was a disaster, losing almost half of its men. The weather had been appalling and the parachute jump risky, but they decided to take the risk; if they didn't, they feared their clandestine outfit would have been disbanded before it even truly got going. Thankfully the second mission was a huge success and the team went on to make a huge contribution to the war effort.

Stirling was eventually captured by the enemy in North Africa, but in little over a year the team had put hundreds of German and Italian aircraft and vehicles out of action from the ground, and had destroyed railway supply lines and communication networks that were crucial to the enemy.

Stirling, like the motley crew he assembled around him, was a maverick. There was little evidence to suggest that what he set out to do would be successful, and without his exceptional levels of tenacity and perseverance, he might never have got it off the ground. But he had a vision that inspired him and he committed to it; he worked hard and he did not allow any initial setbacks to put him off. The end result is one of the most feared and revered elite combat units in the world.

Touched by an angel

I mentioned my friend Ash in Chapter Six – he's the one who threw his shoe at me at university. Well, Ash once told me a theory that struck a chord with me. He said that most people he encountered would rather explain away other people's success with luck, rather than graft – he called it being 'touched by an angel'.

'Touched by an angel?' I asked him.

'Yes. You see, most people out there don't really *want* to acknowledge their potential. They don't want to risk following their dreams, because deep down they are scared of failure. But you know what's even worse? Most people are even more scared of success itself. I mean being *really* successful. Because when you become really successful, then the pressure is always on you to keep being even more successful – and that is terrifying.'

Ash explained, 'It is far easier to give up early on your dreams and blame outside circumstances on not getting there: that way you never have to face the truth. People say things like, "I couldn't do this job that I love because I have a mortgage to pay", or, "I couldn't be with this girl because she's out of my league", or, "I can't go to that country because I don't speak the language", or, "I've never had the opportunity". Or whatever . . .'

It is true. It's much easier to place the blame on something or someone else rather than accept that if you really do want something, and you're prepared to put in the hard work and make sacrifices, you really can do anything you want. It is a big pill to swallow.

So now when I am asked, 'How did you become an explorer?', there are two versions of the truth. The first, much shorter version – the one I tell people if I know that all they are after is a bit of entertainment or an autographed book – is that I was 'touched by an angel'. It goes something along the lines of: 'One day, when I happened to be working in Africa, I got a phone call from a friend of mine who worked in TV, who said, "Do you have any expedition ideas you'd like to pitch?" I told him I'd like to walk the Nile. The TV folks happened to be interested in commissioning an adventure series that year, and well, the rest is history . . .'

Basically, I had a stroke of luck, which meant that the big break I needed came along at the right time – I was 'touched by an angel' and lived happily ever after.

And you know what? People like that story, because it's an easy explanation. They also like it because it's an excuse as to why they haven't done the same things. They feel comfortable in the knowledge that they were not touched by an angel, and therefore they can carry on with their lives complaining about the opportunities they never received.

The much longer version, of course (and I'm not saying this to blow my own trumpet, but rather to demonstrate a point), is that I spent several years travelling around the world in my early twenties with no money, while I was preparing to get in the army. Then I passed the selection to get into the Paras, I fought in Afghanistan and then risked it all to set up an expedition company while sleeping on my mates' floors for three years. This in turn allowed me to establish myself as a photographer and writer, whilst earning enough money to save up to do the expedition I really wanted to do.

I sacrificed my house, my girlfriend and any form of stability to do the one thing that really made my heart sing; that I knew was going to enable me to succeed in what I wanted to do. Yes, of course there were some elements of good fortune and privilege that helped me along the way – it would be churlish to ignore those – but ultimately it was down to good, old-fashioned determination.

But people don't always like that version of events, because it reveals the uncomfortable truth – there are no shortcuts to success.

Resilience and rejection

My dream of becoming a published author was first realised with *Walking the Nile*, but this wasn't the first book I ever wrote, nor perhaps the one I learned the most from.

When I left the army in 2010, around the time that I was setting up Secret Compass, I began writing *Eastern Horizons*. It was the travelogue of a hitch-hiking trip I had done one university summer from England to India, and I wanted to immortalise it and make it the first book I published – all part of the five-year plan that I had worked out with the help of the 'combat estimate'.

I think that first manuscript would be almost unrecognisable to me now. I worked on it for so long, tweaking and editing along the way. I would show it to friends and trusted confidants whose opinion I valued, and I can't imagine how many hundreds of hours of work were poured into that particular book. Once I was happy with the first draft, I started to pitch the manuscript to agents, which is the first step in finding a publisher.

Needless to say, I got rejected by everyone, and to be honest it felt like shit. It's not as if you get all of the rejections in one day, or in one week. Rather, it is a drip feed of misery. A Chinese water torture, where each drip says, 'You're not good enough.'

Even worse than the 'no's are the ghosts. Utter silence – a very singular sort of rejection. What choice do you have, but to pick yourself up and with renewed enthusiasm, charge back into the arena with a load more pitches. No matter the amount of 'no's, I would not be deterred, and *Eastern Horizons* was finally published six years later in 2017. It felt very validating.

When you're up against that level of adversity, you have to dig deep into your reserves to find the tenacity and dogged determination to keep going. The truth is that challenge, adversity and even rejection can be really motivating. After being knocked back by the first few 'thanks but no thanks', I was spurred on by the rejections I was getting from agents. It's a combination of humility and self-belief; I took in the criticism, when any constructive feedback was offered, but I never lost the belief that my book was good enough to be published, to prove those agents wrong.

It's all a matter of perspective. It is the difference between finding out you didn't get a job after an interview, and telling yourself that it was the wrong fit, rather than you were not good enough. Instead of thinking that you will never find the right job, you tell yourself that the right job will come along soon. It is also having the presence of mind to recognise that not getting the job applies only to your work life – it is not an indictment of who you are as a person.

Of course, all of this stuff exists on a spectrum – you might have the ability to be confident and optimistic in one scenario, whilst another totally floors you, but if you can consciously reflect on this difference when you're faced with challenges, you can start to see the world through a more optimistic lens.

Isabella Bird

Isabella Bird was a British explorer, naturalist and writer. Born in 1831, Bird did not have an easy childhood, and suffered from a spinal complaint, insomnia and chronic headaches. Her doctor prescribed Bird an open-air life as her medicine – something that we seem to be coming back to almost two hundred years later.

Ailments aside, Bird became a prominent writer, having her first work published at only sixteen years old. After having a tumour removed from her spine, she later travelled to America, where her letters home formed the basis of more published works. Bird had a thirst for travel, and after America she explored Australia and Hawaii, climbing mountains, and documenting her tales for others. In 1872, she moved to Colorado, which was still very much an unsettled place as part of the new frontier – her close 'acquaintance', 'Rocky Mountain' Jim Nugent was an outlaw, who was shot and killed.

Bird travelled all over the world and became a renowned travel writer, and such was her exploration that she became the first woman to be elected Fellow of the Royal Geographical Society. Life had dealt her a tough hand with her medical conditions, but Bird turned it to her advantage, embracing the outdoors. Through her grit and determination not to be constrained, Isabella Bird blazed a trail for other authors and explorers to follow in her footsteps.

Work hard, work smart

Success doesn't come without sacrifice. I have learned this in so many different contexts in my life: in gruelling boxing sessions throughout my time at Sandhurst; constant drills and training with the Paras; the hard slog it took setting up Secret Compass; and the endless rewrites, edits, rejections and knock-backs it took to become a published author. It is not always obvious to onlookers, who can fail to see or understand the legwork that goes on behind the scenes. The same goes in business as it does on the battlefield. The winners are the ones that make sacrifices, train harder and spend time, energy and effort in preparing to succeed.

There is no way around it, you've got to be willing to put in the hours; diligence goes a very long way indeed. The vision and the dream will not become a reality unless you work for it.

Working hard alone is not good enough; it is also about working smart. Sometimes toil will be necessary, but other times if you take a step back, you can streamline the way you come at things. Discipline is crucial. When I write, I am extremely strict with myself – I plan the structure of the book thoroughly, mapping out how many thousands of words will be in each chapter, and so on. I am always working to a deadline from the publisher, so I have an end date in mind. This allows me to work out how much I will need to write in any given week, or on any given day.

I am often asked if I get writer's block. Of course I do, and unless you give yourself a word count or other requirement per day, it is very hard to overcome that when sitting on your own at a laptop. I always allow a little contingency grace period – a back-up plan of sorts – for those days where I am *really* struggling to express myself or communicate; but fundamentally, if you don't stick to the plan, you are never going to get the thing written. Are you really going to feel more inspired tomorrow, or do you need to be firm with yourself and get started? Getting the words on the page is more than half of the battle, and you can always inspect them with a careful editing eye a few days later.

All that said, spending hours staring at a screen can be debilitating and does not automatically lead to better focus. I delineate time that will be spent on writing, and I turn off the notifications on my computer and phone. I avoid all distractions. When I am really struggling, I go for a walk, or take thirty minutes off and read.

You can also work smart by finding out what your 'frog' is. Mark Twain once said that if you swallow a live frog first thing in the morning, you can go through the day with the satisfaction of knowing it is probably the worst thing that will happen to you all day long. Your 'frog' is your biggest, most important task – the one you are most likely to procrastinate over if you don't do something about it. Once you've done this, you've crossed the biggest hurdle and have achieved something that can launch you forwards through the rest of the day.

Far too many people get caught up in the weeds of detail, embroiling themselves in the minutiae of a plan while ignoring the bigger picture. It's usually because the little wins feel great, but nobody wants to address the elephant in the room. When it comes to expeditions, people start chasing after visas and booking flights before they've got the funding, or even checked to see if a route is viable. Plenty of my colleagues will wrap themselves up in the small joys of spending hours posting pictures on social media, or pontificating over which kind of boots to buy, without giving a thought to the political implications of a journey.

My advice is to think big and work backwards. Prioritise what is important, and what will lead to success. Set yourself goals against timelines and stick to them. Don't worry too much about how you'll get there, because there are many paths to success. Details are important, but they will come later: identify your frog first – then swallow it.

Say No more often

There's an almost cultish predilection these days that insists we must agree to everything. Say Yes! Go and do that 5K run! Go and spend a crazy weekend in Budapest! Go and take up lessons in ancient Sanskrit! . . . the list goes on.

But the truth is, we often go along out of peer pressure, or a misplaced sense that failure to do so will mean we are missing out. What we don't appreciate with this take on life is that by filling our time with stuff and nonsense that we will probably never utilise, or even enjoy, we are wasting precious opportunities to be doing the things that are useful, fun and genuinely fulfilling to us.

As I said at the start of this book, first we need to understand ourselves so that we know what we want, and how we wish to spend our time. The Say Yes More movement is childish and frankly rather patronising. What we should be doing is taking ownership of our lives, and saying No to all the bullshit, so that we can focus on exploring what really makes us happy.

That is not to say don't try new things – that would be entirely against what I talked about in the chapter on curiosity – but only say Yes to those things that make your heart sing, or otherwise might serve a useful purpose. Don't do stuff because everyone else is. If it's not a definite Yes, it's a No.

The key to all of this is discipline. It's about having self-control, and the willpower not to be distracted by all the white noise and temptation that surrounds us on a daily basis.

How is this relevant in the workplace? You can work smart by saying No more. You can't do everything, nor are you good at everything. I'm certainly not – I hate admin and I'm a bit of a Luddite, crap at anything that involves a Google Drive or an iCloud. Be discerning about what you commit to and then do it at full pelt, instead of spreading yourself too thinly and burning out.

Remember to factor in your body when you are trying to understand your own behaviour. Often our behaviour is dictated by physiology; say lack of sleep or low blood sugar. It is critical to be aware of your own energy levels and work out what is called your biological primetime; knowing what drains you and what energises you. Do you get your energy from being in your own company, or being around others? What boosts your mood and what drains you, and when are you at your most energetic? Are you a night owl or a morning person?

Knowing when you function best, your biological primetime, can be a great help. Record your personal energy throughout the day, noting what times and what things energise you. Once you've established this, you have greater potential for productivity and can plan your work and life around these ebbs and flows. If you're an extrovert, you may schedule meetings at a time when you're usually flagging, as they'll give you energy. If you're an introvert, this is more likely to be something that costs you energy. If you are able to concentrate all morning, but have an urge to snooze after lunch, schedule that time to do something mundane or non-creative, like replying to emails.

Discipline, graft and determination are critical to getting stuff done. Understanding your own needs, and what makes you tick, will make you more efficient, and finally accepting what is and isn't within your control will help you to navigate the complexities of any situation. What comes next is perhaps the hardest task of all – defeating your own ego and mastering yourself.

Palmyra, Syria

10

Admit When You Are Lost

Humility is not thinking less of yourself, but thinking of yourself less

Among the ruins

There are many treasures around the world. Many are natural and seemingly impervious to disasters, others man-made and artificial. Those, like the ancient city of Palmyra, have stood for thousands of years with no guarantee of their permanency. Palmyra, one of the best-preserved Roman sites in the world, resides in what is now the bloody battleground of Syria. I had always wanted to visit its ancient Semitic palaces, and now that I'd trekked around the Arabian peninsula all the way from Iraq, I had my chance.

The problem was that Palmyra was still very much on the frontline against ISIS. It was 2018, and they were on the back foot, but there was a genuine threat of attack and a grisly death. I couldn't help myself, though. I had to see it, and I was rewarded in this decision when the old town rose from the sandy plains like a mirage. Considering that it had twice been taken by ISIS, I didn't expect to see much more than a crumbled reminder of what once stood there, and a sense of relief washed over me when I saw how much of it still remained.

As I wandered between the ruins, I was overcome by a mixture of emotions. The damage was clear to see. Gone was the Triumphal Arch, the gateway to the Roman city. The Temple of Baalshamin lay in sad wreckage, along with much of the Great Colonnade. Though I was glad to see that many of the columns still stood strong, and that the Baths of Diocletian, and the Senate, were for the most part undamaged, I felt a strong sense of smallness, of the transitory, trivial nature of my problems in relation to the scale of these buildings, and of the world and time itself.

I looked up in wonder at the infamous Roman Theatre, which once held gladiatorial contests. In equally grim circumstances more recently, the executions of hundreds of men by Islamic State terrorists were held here. The Temple of Baal was utterly demolished, blown up by these modern-day philistines, and now only the central gateway remained. It was here, standing on top of the rubble of a temple to an ancient God, where I met Tarik Al-Assad.

'Tarik's father was the head of antiquities at Palmyra Museum,' Nada, my guide, introduced us. 'He has quite the story to tell.' He held out his hand for me to shake. I studied his face as we greeted one another, guessing him to be about my age. A sad look in his eyes betrayed an unknown tragedy.

'I work at the museum in Damascus,' he said, leading me through the piles of 2000-year-old masonry. 'My father was in charge here. His name was Khaled, please do not forget it.'

I nodded, willing the story to continue.

'My father loved this place, he worked here for forty years. He was a historian and a hero. It was because of him that Palmyra became a UNESCO world heritage site. When Daesh came here and took over, they arrested him, and me too. For weeks before, we had been taking the antiquities to a place in Damascus for safe keeping, because we knew that Daesh would steal them and sell them in Turkey. Before they came, my father told me to run away with my mother and brother, so we escaped one night and hid from them. But my father refused to leave.'

Daesh was what the locals called ISIS. I swallowed, knowing how this story would likely end.

Tarik paused as though steeling himself before continuing. 'He could have left with us, but he said his place was in Palmyra, and if that meant going down with it, so be it. Daesh were stupid and thought that my father was hiding gold in the ruins. They beat him and asked him where it was. They didn't believe that there was nothing, so they began blowing things up. When they realised it was just old stones, they demanded that he reveal the antiquities. My father refused to tell them anything.'

Tarik stuttered his words and I noticed a tear running down his cheek.

'Then one day, on the eighteenth of August 2015, they took him to the crossroads by the mosque, just over there.' He pointed beyond the gate to the shattered remnants of the modern town, where a bullet-strewn minaret jutted miserably into a grey sky.

'They forced him to kneel down on the pavement, and cut his head off with a sword.'

He fell silent and gazed off into the distance.

'He was eighty-three years old,' Nada added beside me.

'Look at this!' Tarik exclaimed suddenly, pulling his phone from his pocket and forcing it into my hand. I peered at the screen in disbelief, looking at an image so gruesome I could barely look for more than a couple of seconds. A body was hanging upside down from a traffic light, its decapitated head sat in a pool of blood below on the pavement, still wearing a pair of broken spectacles.

'It's my father,' whispered Tarik, forcing a smile. 'I still have his glasses.'

In a state of shock, I didn't know what to say. Tarik's father had been murdered and the criminals who had done it had sent this photo to him to torture the family even more. I asked Tarik what he would do if he ever met his father's killers.

He smiled. 'It'd not for me to punish them. That is God's decision. I must forgive them.'

Tarik had no anger and he had no urge for revenge, and he was putting his hand on my shoulder, as if *I* were the one in need of comfort. It was a reminder of the greatest of all virtues – *compassion*.

The Golden Temple

In the heart of Amritsar, in the Indian state of the Punjab, lies the Golden Temple, the most holy shrine in the Sikh religion. At its centre is a shining golden shrine surrounded by a placid square of water, in which the faithful bathe to purify their bodies and their souls.

Around this is a walkway on which Sikhs perambulate or prostrate themselves before the sacred temple, and the air shimmers with the soothing sounds of sermons broadcast in Punjabi through speakers at the four corners of the space. Its four surrounding walls face North, South, East and West as a symbol that its grounds are open for all races and religions of the world to enter, and its many kitchens serve food for free to ten thousand people, regardless of religion or status, every day.

To enter the temple complex, visitors must cover their heads as a sign of respect, and walk through small trickles of cleansing water to purify their bare feet. Then, they descend a small flight of stairs, a symbolic recognition of the humbleness of all people before God.

Guru Nanak, the founding father of Sikhism, taught that humility is key to fulfilment. Sikhism's holy book, the Sri Guru Granth Sahib, describes him as 'the companion of the lowest of the low and of the condemned lot. He has nothing in common with the high born'. According to Sikhism there is no place for Ego, or Haumain, in the sphere of Divine Love, and in the House of Guru Nanak Garibi, Nimrata, (Poverty, Alleviation) and Humility reign supreme. Because all people are equal before God, there should be no hierarchy, only humility.

Unlike some other religions that claim a monopoly over truth and wisdom, Sikhism recognises the ability of all other religions and faiths to enlighten their followers. In this sense, the quiet, reflective tranquility of the Golden Temple is one of the most humbling places on Earth.

Humility

Sikhism is not the only world religion to espouse the virtues of humility, though perhaps you could argue that it goes about it in one of the humblest ways (the gold sheet-covered temple at its heart notwithstanding). Philosophers and spiritual leaders across the world and throughout history have recognised the importance of humility in understanding our place in the world, and leading fulfilled and purposeful lives. Jesus taught 'Blessed are the meek', and the word Islam itself translates to 'Surrender (before God)'. Buddhism and Hinduism teach that humility and self-improvement are key steps on the path towards higher states of being.

What many religions teach is that humility is not the same thing as self-deprecation. Self-deprecation is a form of arrogance, a focus on yourself and a comparison between your own poor circumstances and an imagined better state, of which you are implicitly more deserving. Or in the words often wrongly attributed to C.S. Lewis, 'Humility is not thinking less of yourself, but thinking of yourself less.'

Humble leaders have been shown to have a huge impact on business performance. Researcher and consultant Jim C. Collins led a 1996 study that sought to identify the common factors between companies that went from a 'good' performance (i.e. broadly in line with market average for several years) to a 'great' performance (share price three or four times the market average for a sustained period).

The most significant factor that emerged from their four-year project was what Collins calls 'Level 5 Leadership' – and their definition of 'Level 5 Leadership' is counterintuitive, even by Collins' own admission. For all we've come to associate the vision of successful business leadership with big, flashy personalities like Alan Sugar and Richard Branson, the most effective leaders identified by this study were characterised by 'a paradoxical combination of personal humility plus professional will'.

These leaders themselves, when interviewed as part of the study, showed an unwillingness to credit their organisation's success to their own actions, downplaying their own contributions and up-selling the strength of their teams. One of the key criteria for a 'good to great' company in this study was sustained success. It wasn't enough to improve a company's fortunes temporarily; it had to show sustained, three-times-the-market results for fifteen years.

Collins points out that a lot of the companies that were discounted from their 'good-to-great' categorisation for this reason happened to have very arrogant, showy leaders. They might have temporarily turned a company's fortunes around, but soon became so absorbed in cultivating their own personal brand and self-promotion that company performance slumped.

To achieve anything great, it is important to recognise our own small, humble place in the world.

Inspiration all around

When I walked across the Sahara desert in 2014 as part of my expedition to walk the Nile, I found myself in a bit of quandary. I had reached my mental and physical limits, having already walked for more than six months over some of the harshest terrain in Africa, but nothing could have prepared me for what was to come. The searing heat of the desert, the empty barren wilderness and seeing nothing but sand for days on end. On top of that I'd got terrible blisters, and every inch of my body was in constant pain from the ceaseless walking.

I knew that I couldn't travel this part alone and would need some support. It came in the form of my Nubian guide Moez, and two Bedouin camel drivers, Ahmed and Ahmad, plus of course their three camels, which I acquired in Khartoum.

I worked out my progress and figured that we needed to reach the Egyptian border, some hundreds of miles to the north, within two months. The reason was that as soon as the Islamic month of Ramadan had begun, the border would be closed, and I would be stuck in Sudan for thirty days longer than I needed to be. Moez and the camel drivers agreed – in fact, the Bedouin said that we had sixty days and no more, because wherever we were then, they would turn back to make sure they had enough time to return to their families for Ramadan.

So that was it, decided. We trundled north into the Sahara, each day more arduous than the last. For weeks on end we would walk at least the distance of a marathon every day, sometimes up to fifty kilometres. It was the most punishing and demanding thing I had ever done. I couldn't have been good company: I was tired, lonely and fed up of the whole thing. I spent half my time trying to catch a mobile phone signal, so I could keep up with the outside world and contact my unresponsive girlfriend of the time. I was thoroughly miserable and wanted to get to the end as soon as possible.

To try to keep the logistics as simple as possible, I had planned to stay close to the River Nile, where it would be easy to obtain water, and when we passed by villages we could buy food and supplies.

The only problem was that Ahmed and Ahmad didn't like walking next to the river. 'We are Bedouin – people of the desert!' they shouted, theatrically. 'The camels prefer the sand.' As it turned out, my hardy camel men were simply scared of snakes and mosquitoes.

A bigger issue was keeping to the schedule when faced with overwhelming hospitality. Despite what I'd read in the media about Sudan, with its tribal violence, poverty and dictatorial government, its people are some of the most welcoming I have ever met. All along the famous river, I was greeted with a friendly smile and offers of endless tea, delicious flatbread and free meals of goat stew. It was hard to say no. The villagers never wanted anything in return, except the honour of hosting a foreign guest for the night.

For me, after the horrors of travelling through war-ravaged South Sudan, it came as a welcome break and soon helped me rise out of my depressed state of mind. However, it was less welcome for the other team mates, who were more concerned with making sure they got home in time for Ramadan. 'Ah, this is just normal here,' they said nonchalantly. 'We are used to it.'

For me, though, it was truly humbling. People often ask how I managed to stay motivated during the toughest times of my walks, and I often think of those Sudanese villagers. Even though they lived in abject poverty and had nothing of material worth, they would rather have died themselves than deny a drink of water to a guest. I reminded myself daily that no matter how lonely or demoralised or tired or hot I was, it was only temporary. These people lived their whole lives here in these desperate conditions. It was that incredible sense of generosity that got me through – those people with their kind smiles.

They would run after me laden with sweets or drinks, insisting that I stay a little while longer to enjoy their hospitality. Ahmed and Ahmad would always try to shoo them off and explain to me that we must hurry along. I even got to the point where they threatened to mutiny unless we avoided the villages entirely and followed the desert paths instead. In the end we reached a compromise, whereby every other day we would walk a mile or so away from the river to avoid the settlements, and thus risk being slowed down by the offers of lunch.

On many occasions it would prove fruitless anyway. Even a mile away from the villages, whenever we made camp and Moez lit the fire, dozens of villagers, intrigued as to why there would be a fire in the desert, would come wandering out to the dunes where we sat on rugs under the stars.

'What are you doing here?' they would shout.

Moez, or the camel men, would try to explain the situation, but the villagers were always upset that we would not accept their hospitality. One local man was so horrified that he stormed off into the night, and returned half an hour later, having carried his bed on his head.

'Here,' he huffed. 'If you will not come into my home . . . my home will come to you.' He slammed it down into the sand and bowed gracefully.

I wondered if I would ever be greeted with that sort of response back home in England. I seriously doubted it.

I swore I'd never complain about my lot again.

'The only thing we know for certain, is that we know nothing'

. . . so said Socrates. According to Ryan Holiday, author of the book *Ego is the Enemy*, many of our problems and challenges in life are not the fault of external factors, but are born of our own ego – an arrogant, 'self-centred belief in our own importance'. There are three phases we go through in which our ego plays a role, and our ability or otherwise to master it can determine success or failure, happiness or unhappiness.

Firstly, Aspiration, the desire to accomplish bigger and better things in life; secondly, Success, when we achieve these goals and receive praise for it; thirdly, Failure, when we don't achieve our aspirations and the consequences of this.

Aspiration is driven by our ego, and in that sense our ego serves a positive purpose. But when we start imagining and talking about our future accomplishments, it can trick our ego into imagining these things have already happened, and we can forget to put in the necessary work. It breeds the worst forms of arrogant behaviour. We focus more on the material perks of success, rather than the impacts that our actions have on the world. We become lazy, forgetting to put in the work. We convince ourselves that we have everything figured out, and stop learning from new experiences.

Success is one of the most dangerous threats to the ego. There are endless accounts of people who work extremely hard to achieve a position of power and influence, only to throw it away from their own hubris. It's an intoxicating drug that robs us of the sobriety required to sustain it. Failure hurts the ego. It's a direct challenge to our misplaced view of ourselves as the centre point of the universe. Our ego amplifies the negative aspects of failure, sending us into a downward spiral, where failure is seen as the end of the world, rather than an opportunity to learn and grow.

This all runs completely counter to the explorer's mindset. Our egos, if left unchecked, have the potential to stifle every aspect of our journey that makes the world worth living in. We lose sight of the self-knowledge and self-improvement that are key to a fulfilling existence; we succumb to fear because we lose our curiosity, and thereby our optimism, about the world around us. We alienate others, losing sight of the virtues and morals that have guided our decisions, and become poor leaders; we lose the capacity for diligence and hard work that are crucial to success. Without humility, we are unable to explore. Sometimes it's better to accept you are wrong, even if you think you are right, as difficult as that might sound.

The Darien Gap is one of the most remote and inaccessible jungles in the world. It is a stretch of rainforest in the lawless no-man's-land between Panama and Colombia that separates North and South America, and even now is a byword for danger amongst the hardiest of explorers. It is the kind of place where anything is possible.

I was trying to cross through the notorious jungle back in 2016, as part of an expedition to walk all of Central America. I'd left the highway behind and started upriver by dugout canoe, which for much of the time we had to push up cascades and around fallen trees. I had been to a lot of remote jungles in my time, but this was another level. The only way to navigate through this treacherous terrain was to enlist the help of the local tribe – the Embera-Wounan people – and so it was that I found myself in the company of a dozen or so of these men armed with machetes and bows.

After two days struggling upriver, we came to the final settlement before the mountains, and it was where most of the men came from. They insisted we spent the night in the village, where we were given a hut on stilts to protect against snakes and flooding.

As night fell, we were treated to a meal of roasted fish and rice by the chief's wife, who came and sat down next to me. Maria might have looked old and withered, but she had a twinkle in her eye that betrayed a lifetime of mischief, and she was most definitely the one who wore the trousers.

'Do you want to buy something?' she said in a conspiratorial tone.

'What?' I asked, expecting to be peddled some tribal beads or needlework, like we'd seen in the markets of Panama City.

Instead she pulled out of her apron a cloth, which she unwrapped to reveal a white thing the size of a thumb. It was a tooth.

'It's from a Jaguar,' she said.

I grimaced. *Poor thing*, I thought. *They've probably killed it to sell the teeth*.

'No thanks,' I said, explaining that I didn't want to encourage poaching.

She thought about it and shrugged her shoulders, wrapping it back up in the cloth.

'Okay, I have something else,' she said with a sigh. 'I've had it a very long time, but I want to sell it.'

From another pocket came another cloth, this time bigger. Before she unwrapped it, she looked me in the eye and stared.

'These are very rare,' she said in a whisper.

'Let me see.'

She placed the object into my hands and I felt the heavy coldness of something hard and triangular in shape. In the half-light of the room, I couldn't tell what it was. Two of the sides were razor-sharp and serrated, and it looked like some type of Stone Age spearhead made of flint. That must be what it was, evidence of a prehistoric society in Panama perhaps?

'No, it's not a spear,' said Maria. 'It's a *Una de Raya*.'

'She says it's a nail of lightning, like a fingernail,' said my guide. 'I think she means it's a thunderbolt.'

I laughed out loud. She frowned and grasped it out of my hand.

'It is, it is, that's what it is. My grandmother found one. They are only found in the Darien, nowhere else. We sometimes find them underneath trees that have been burned by the sky.'

I supressed my chuckles and felt bad that I had laughed. Maria genuinely thought that it was the tip of some sort of supernatural thunderbolt, presumably petrified upon impact with the earth.

If it wasn't a spearhead (it was too big and irregular for that) and it wasn't a thunderbolt, there was only one thing left that it could be. It was a shark's tooth. But this was not any old shark's tooth, it was that of a prehistoric megalodon; a thirty-metre monster ten times bigger than your average Great White, which swam in the oceans over three million years ago. One thing was certain: this ancient fossil was formed at a time when Central America was nothing but a ridge of underwater volcanos, but then again, who was I to argue with this wise lady and her dreams of thunderbolts?

Even if I had a notion to correct her, what is a more likely story: a flash of lightning striking a tree and leaving a piece of stone in the shape of an arrow, or a giant sea monster leaving its teeth in the middle of a jungle? The old lady had never even seen the sea, nor had any concept that her piece of forest was once under it. Sometimes it is wise to hold your tongue, because what is true for one, might not be true for another. Choose your battles wisely.

So I told Maria that I'd love to buy her thunderbolt, even if it was a hundred dollars. I didn't barter.

Heinrich Harrer

Born in 1912 and raised in the Alps, explorer Heinrich Harrer grew up utterly obsessed with climbing and skiing. He was gung-ho about adventure, once surviving a 170-foot cliff fall. In the mountaineering world, he is famed for successfully scaling the almost vertical north face of the Eiger mountain in the Alps. But it is not his climbing exploits on which I wish to focus, but his unlikely friendship with a young boy.

In 1939, Harrer was in the Himalayas planning an ascent on Nanga Parbhat, when war broke out, and as an Austrian he was interned in British India in a prisoner-of-war camp. After several attempts, he and a companion escaped and hiked out across the mountains. They were looked after by nomads and treated with kindness and hospitality, and after almost two years of trekking in freezing conditions, and avoiding bears and roadside robbers, they made it to the city of Lhasa. Bear in mind, at this time, Tibet was resolutely closed, letting no one through its borders. But by some miracle of hardiness and persuasion, they got to the forbidden city.

It would turn out that, in reality, the Tibetans were curious and inquisitive about these two foreigners and, above all, they showed them compassion. Despite being ordered to leave, they sat tight and an aristo-crat took pity on them, letting them kip in his garden and feeding them. They got to work making themselves useful and soon the Tibetan government had given Harrer a job as a translator of foreign news and an engineer, and they gave him a home and servants. Part of his work also included cinematography, helping them with photography and film.

Here in Lhasa, they soon realised that the entire city revolved around a ten-year-old boy. His holiness had been discovered as the 14th Dalai Lama at just two years old, and as the human reincarnation of God was both Buddhist spiritual leader and Tibetan king. At his request Harrer was recruited to be the god king's tutor.

The young Dalai Lama clearly had a hunger for learning, and was diligent enough to do double the translation homework that Harrer would set him. He was curious about the mysteries of the outside world; he'd heard about aeroplanes and wanted to know how the engines worked. He would gobble up news and knowledge about the technical advancements happening in the Western world, even though they must have seemed utterly alien. He would ask about global politics and was particularly interested in geography and maps of the world, working out where Tibet fitted into it all. With twenty-three years between them, they were a rather unlikely pair, but this didn't stop them from talking for hours about Buddhist philosophy and Western science.

Ever resourceful, Harrer would keep the king entertained, running a cinema projector of a jeep engine – one of only a few cars in the whole region. I'm sure it made a welcome break from all the priests and reverence and rigmarole – underneath it all, of course, he was also a teenage boy. Besides, he had very few friends his own age, a strict study regime and was likely desperate for stimulation, variety and companionship.

Chinese troops invaded Tibet in 1951, forcing Harrer to flee to India. Not long later, and far more famously, the Dalai Lama was forced to flee himself. He has never been back and has lived a life in exile, having sought refuge in India. The Dalai Lama would go on to become a global leader, both as the spiritual head of Buddhism and as the political leader of 100,000 or so Tibetans in exile. I was lucky enough to meet him, getting a private audience with him in Mcleod

Ganj, his home in exile in the Indian Himalayas. Now eighty-four, he campaigns internationally for his exiled homeland and travels the world sharing Buddhist teachings and words of wisdom.

Harrer went on to write a book about his time in the Himalayas, Seven Years in Tibet. Published in the 1950s, it was the first time the Western world had heard any stories from this secretive kingdom. Even at a tender age, the Dalai Lama made a lasting impression on Harrer; the tutoring did not go only one way. Heinrich Harrer learned a huge amount from the boy.

He learnt the Buddhist philosophy of tolerance and about the belief of peaceful action; he was struck by the fact that Buddhism never sought to convert non-believers. He learnt about the idea that wealth is impermanent and will not grant you happiness, and the idea that we need to live each day as it comes, instead of dwelling in the past or focusing too much on the imagined future. He never carried a gun again, based on the non-violent Buddhist teachings he'd learnt from the Dalai Lama.

Despite all the trappings of power and wealth, Harrer described the young Dalai Lama as unspoilt and humble. He chose to fast and keep silence, keeping an ascetic way of life. There is a lot that we can learn and adopt from the Dalai Lama before he even grew into a man. In Tibet, Heinrich Harrer would find a sense of peace and self-awareness that he summed up thus: 'I have learned to contemplate the events of life with tranquillity and not let myself be flung to and fro by circumstances in a sea of doubt.'

Life of learning

For me, the most important thing about humility is that it reminds us of the limits of our own knowledge, which gives us curiosity and a desire to keep learning about the world. One of the most useful insights I have gained in life is that success is best measured against yourself. It's not profound, it makes perfect sense. But it's not something that we teach in schools, where we tend to pit children against each other, feeding the more negative sides of their egos, and creating the impression that success means doing better than other people.

It's not something that comes naturally to me either. I have quite a competitive streak and relished the rivalry as a boy. It can be a long time before we figure out that it doesn't matter what all of your peers are doing. What this revelation looks like for me, is a lifetime of learning and self-improvement. Like many children, I was hungry for learning and knowledge, but in our society this seems to dissipate and fall away as we get older. When school and university come to an end, we are sent out into the world, and for lots of us this stimulus and sense of constant personal development disappears.

In the early part of my career, in the army, I was lucky enough to be continually learning. But it was when I left and started out on my own that I understood the onus was on me to go out and learn new things. I had always read a lot and had a healthy curiosity, but this was the first time I'd really ventured into autodidacticism: self-education without the guidance of teaching or an institution.

Some of the most useful and important things that I know now, I had to learn on my own: from pension schemes and tax returns, to relationships and learning to communicate how you feel. More importantly now, because of the internet there is far too much knowledge for us to possibly absorb – and this is only getting harder. Everything that I learnt in Year 5 history, I could digest from one Wikipedia page. We are too focused on information.

One of the things that gives us the edge over animals is our ability to think collectively in large groups, but more and more we are becoming reliant upon the knowledge of others to create our understanding of the world. There are huge benefits to this collaboration, but there are also pitfalls. We think we know far more than we do. It's all an illusion. We treat knowledge in the minds of others as if it were our own, when we don't know how many of the simple things around us work. It's easy to think this comes from inside our own brains, when really it is drawn from the world around us and from our community.

Instead of our endless obsession with knowledge, we need to focus on teaching children the real-life skills of critical thinking, collaboration, communication and creativity. We also need to teach children the ability to embrace change, which is a lot harder than any chemical equation. This means giving kids a mindset that teaches them they can continually improve. In one study, children that pushed through their challenges were exclusively the ones who believed they could improve their own abilities; their trajectory was not fixed and they could grow. Those who pulled back, afraid of the challenges and not going for it, were the ones who held the belief that their ability was fixed.

It is the difference between the two outlooks, 'I didn't do very well in that maths test' and 'I'm bad at maths'. Some schools have even been giving kids the grade 'not yet'. This kind of thinking encourages young people to keep learning, to recognise that however much they already know, there is a lot more to learn out there, and it cuts out many of the dangers that can come from over-inflated egos.

Captain Tom Moore

By April 2020, you could argue that Captain Tom Moore had done his bit, his fighting days long behind him. A Second World War veteran, who had served with the Duke of Wellington's regiment in India, Sumatra and Myanmar, he was approaching his 100th birthday when the world was plunged into lockdown in response to the COVID-19 crisis.

'Captain Tom', as he has become known, could have been forgiven for sitting out the latest 'war' (as he described it), considering his age and relative frailty. Instead, he set about fundraising for the NHS Charities Together foundation, famously completing 100 laps of his garden with the assistance of a walking frame, ten 25-metre laps per day, before his centenary.

His initial fundraising target of £1,000 was smashed; the final total of £32,796,475 is a record for any JustGiving campaign. It captured hearts and minds across the world, and led to Moore being Number One in the UK charts for his birthday after recording a charity single with Michael Ball. It wasn't a desire for fame that had motivated Captain

Tom, but a view that however small a contribution he might make, it was worth trying to do, to make an impact on a cause bigger than himself.

Captain Tom's humility was a big part of the reason why he was successful in his goal, and why people across the world bought into his story. As the difference between his original target and the amount he finally raised shows, it is worth doing something for a worthy cause that's bigger than you, and we always underestimate the impact our actions can have when we put our ego aside in the face of a higher cause. Every time we do so, we're surprised and humbled by the results.

Everything in perspective

How can we learn to control our egos, and become humbler and more fulfilled in the process?

I could try to sum this up myself, but it would be a vain task indeed when Rudyard Kipling has already expressed it so eloquently in his poem, 'If':

> *If you can keep your head when all about you*
> *Are losing theirs and blaming it on you,*
> *If you can trust yourself when all men doubt you,*
> *But make allowance for their doubting too;*
> *If you can wait and not be tired by waiting,*
> *Or being lied about, don't deal in lies,*
> *Or being hated, don't give way to hating,*
> *And yet don't look too good, nor talk too wise:*
>
> *If you can dream – and not make dreams your master;*
> *If you can think – and not make thoughts your aim;*
> *If you can meet with Triumph and Disaster*
> *And treat those two impostors just the same;*
> *If you can bear to hear the truth you've spoken*
> *Twisted by knaves to make a trap for fools,*
> *Or watch the things you gave your life to, broken,*
> *And stoop and build 'em up with worn-out tools:*

If you can make one heap of all your winnings
And risk it on one turn of pitch-and-toss,
And lose, and start again at your beginnings
And never breathe a word about your loss;
If you can force your heart and nerve and sinew
To serve your turn long after they are gone,
And so hold on when there is nothing in you
Except the Will which says to them: 'Hold on!'

If you can talk with crowds and keep your virtue,
Or walk with Kings – nor lose the common touch,
If neither foes nor loving friends can hurt you,
If all men count with you, but none too much;
If you can fill the unforgiving minute
With sixty seconds' worth of distance run,
Yours is the Earth and everything that's in it,
And – which is more – you'll be a Man, my son!

Ultimately, we need to create a world in which we all recognise that what we currently know is not the extent of what is possible, and that we can continue to enrich ourselves and one another if we adopt the humility to listen and learn.

Our own egos can be the greatest demons we face. In this sense, they're a little like Mark Twain's frog – once we have overcome them, we are ready to find fulfilment within ourselves.

Jacques-Yves Cousteau

II

Conquer Your Own Everest

*We shall not cease from exploration
And the end of all our exploring
Will be to arrive where we started
And know the place for the first time.* – T.S. Eliot

Look after yourself

You know when they tell you on aeroplanes to always fit your own mask before helping others? The reason of course is that you can't be much use to anyone else if you're conked out unconscious. The same goes for happiness. If you're sad and miserable nobody is going to take lessons from you on how to live their lives. You can't go around preaching if you haven't got your own house in order first. That's why it's important to focus on yourself and your own internal journey first. Those in glass houses and all that.

That's not to say to ignore others, act selfishly and never extend a helping hand – of course not. But don't be a martyr either. When you feel fulfilled, you have a lot more to offer those around you, like a cup overflowing with water. Be a full, happy, overflowing vessel and you'll explore places you never imagined possible.

Take the time to learn your craft, become good at what you do, and do what you love. Life is way too short to be doing things that make you unhappy, so stop wasting time putting things off and focus on the task in hand. That's the first step to self-mastery and the first step to the art of exploration.

Enjoy the journey

In 2019, I decided I needed to be in one place in order to relax. I was feeling burnt out having been on the road pretty much constantly for a decade and I felt like all my recent travels had been for work, leaving little time to just appreciate what was going on around me. As I looked for somewhere to take a break from expeditions, Bali jumped out as an eminently sensible place to stay put for a few weeks. I had never been to Indonesia and I'd heard the food was good, it was relatively inexpensive, and I quite fancied getting into a routine that involved healthy eating and yoga.

As I touched down at Denpasar airport among throngs of Australian backpackers, I realised that Bali was the 100th country that I'd visited. I am not much into making bucket lists or counting where I've been, but it was a question I kept being asked, and one day I sat down with a map and figured it out. Bali was as good a place as any to celebrate half a lifetime on the road, so I was quite looking forward to a couple of days relaxing on the beach with a cocktail or two to get into the swing of things.

That being the case, I was rather disappointed when upon arrival at my little villa in the resort of Canggu, I was informed that the next day was Nyepi, a Balinese public holiday known as the Day of Silence.

Bugger.

Nyepi sees Hindus across the island practise fasting and meditation. Strict restrictions against working, entertainment and even leaving the house are in place. Electricity is typically turned off, and there is no internet or even phone signal. Nyepi is the Balinese New Year, the day when Hindus contemplate what has gone before and cleanse themselves of past sins. The silence and self-imposed exile is supposed to free the island of demons and devils, because when they come out at night to exercise their evil, they find nobody in the streets so they give up and leave, looking elsewhere for their victims. On a spiritual level, it's the equivalent of a reset button where people pray and look forward to a more hopeful future.

Locals don't work, make noise, play music or watch TV, and if you are doing it properly, you aren't even allowed to eat during Nyepi. It's a day where people simply stay at home and meditate. No one is exempt, not even tourists – who are forbidden from leaving their hotels or villas. Bear in mind this was in the days before the COVID pandemic, when all that became a norm . . .

Despite my initial reservations about a potential spoiled holiday, it turned out to be just what the doctor ordered. As I found out, it was rather nice being without the internet or electricity, and the fact that I couldn't leave the house meant that I sat around in silence, reading and thinking. It was actually pleasant that I couldn't do any work, and the fact that I was a prisoner to this cultural phenomenon meant that I felt an unspoken, yet incredibly powerful connection with all the other people who were having the same experience.

For days after, people asked each other how was their Nyepi. Did they feel okay? Did they feel renewed or reborn? I won't go so far as to say that I felt reborn after a day of no WhatsApp, but I could see the appeal of this communal purge. It also gave me time to grasp the value of enforced rest and contemplation.

It was the first time in any culture where I had been effectively 'forced' into joining in a religious practice. Even in strictest Afghanistan and Iran during Ramadan, I had been given sympathetic looks and the sly offering of a loaf of bread under the table. It's not often we are obliged to consider our beliefs under such duress, albeit in the tropical surroundings of a beautiful island where it's not such a drag. For most, it is tough to have to observe another person's customs, or even at times, one's own.

I imagined the uproar if this sort of spiritual incarceration was forced upon the people back home. There would be outrage. I asked the housekeeper, 'What if someone is ill or about to give birth and needs to go to hospital?'

'Of course, for that, we do make an exception for emergencies. But you must understand,' he went on with a smile, 'nobody gets ill or has babies on Nyepi, that would be inconsiderate.'

While I was in Bali, I spent every morning practising my downward dog yoga exercise and ate lentils. There is a sense of satisfaction that comes with gradually building your strength back up and getting fit again, and the bodily enrichment that comes with that; there is something very nourishing about getting into a slow, physical routine where choice is limited.

I had found that was beneficial on my long walking expeditions too. You go to sleep feeling physically exhausted, but deeply satisfied at the end of the day – you wake with the sun, sleep at dusk and eat whatever is presented to you. When there is no phone signal, or distractions from technology, the body will naturally relax into a more natural evolutionary pattern, refreshed and ready to make the most of another day.

I think we can all benefit from a little Nyepi in our lives. It might not go down well if you cut the power to your neighborhood, but maybe try at least a little time in your life where you turn off the devices, stay away from the screens and give yourself the silence to really *think* without the noise. I hope that you enjoy it as much as I did.

The Turning of the Bones

Philosophers have long recommended the ancient method of memento mori to focus the mind on the present. Memento mori means to keep a symbolic reminder of the inevitability of death as a way to encourage people to live while they can, and not to waste precious time on pointless worries. Usually it comes in the form of an emblem – a piece of art, a talisman or photograph. Christianity uses the crucifix, and Tibetan Buddhists used to keep images of skulls.

On the isolated island of Madagascar live a tribe known as the Malagasy, who have an unusual relationship with death. Every five to seven years, they dig up their ancestors. This unusual family reunion is known as Famadihana or The Turning of the Bones ceremony.

Family members carefully remove their loved ones from crypts and burial grounds. They take off the garments and cloth around the bodies. They carefully re-dress the bones in fresh linen before they begin to celebrate and pass around the bones for others to inspect. In their worldview, you haven't left the earth and died until you have completely decomposed. The family members make blessings and offerings and share stories of their deceased. Those who have died are seen as between the living and God, and have some power over the events and activities on earth.

The ceremony is seen as a joyful and happy time. It is more of a family gathering, with a feast and party atmosphere. But in this family reunion, the dead are also invited along. Guests can travel for two days to reach one of the ceremonies, bringing offerings of money and alcohol along with them. The bones are then returned to the tomb before sunset. The ceremonies act as a communication between families and ancestors and a reminder of death. This ceremony and celebration of death is a reminder to those still living to enjoy life and live in the moment while they can.

The only certainty in life is that it is a mortal game. We're all going to die eventually, and so what really matters is what happens in the meantime, and I'd like to think that we have at least some control over that. Being human is all about choice, and how you choose to live is ultimately up to you. Taking control of your own choices is perhaps one of life's greatest challenges, yet when we do, it can be our greatest reward.

Finding meaning

In an age where we seem to be more and more disconnected from our environment it's vitally important that we look after our minds and bodies. There are many ways in which we can do this, through physical fitness, mindfulness and meditation. One of the most useful things that I have learnt on my travels is to consciously foster a sense of empathy with those people that I meet, and it's something that I try to do at home and in my daily interactions too.

Empathy is what fuels connection, and human connection is what brings meaning to our lives. This ability to see things and understand things from someone else's point of view, without judgement, is what gives us purpose and creates a sense of belonging.

In her maiden speech to parliament in 2015, the late Jo Cox MP said, 'We have far more in common than that which divides us.' It is a good message to heed. Some people pray, others fast or read books on enlightenment, and yet more form communities or have children as a way of giving meaning to their lives.

I've visited myriad cultures around the world, from the cannibalistic Aghori monks of India to the polygamous Muslim tribes of Central Asia, and on the face of it they are all very different. Ostensibly, the long-neck Karen women of northern Thailand don't have a great deal in common with the Kuna tribe of southern Panama; nor do the cattle-herding Mundari have much similarity to their Nubian desert-dwelling neighbours of Sudan. The blue-eyed shepherds of Pakistan's Hunza valley couldn't appear more different to the black-skinned Nilotes of northern Uganda, and if you put a Yakutian bear hunter next to a Saudi Arabian mullah, they might struggle to chat.

And yet, wherever I have been in the world, I have discovered that people are more or less after the same thing. Despite surface differences in religion, faith, codes of practice and ethics, generally speaking people just want to have a sense of belonging: to love and be loved.

Trust the universe

It is also important to have a little faith. I don't necessarily mean religion, or spirituality, although the end goal is ultimately the same. Having faith in the universe, or if you will, faith in your own place within it, is the key to acceptance – after all, we're all part of the same spinning rock. Frank Lloyd Wright famously said, 'I believe in God, only I spell it nature.'

We're all in a constant battle with time. Some people worry about the past and get depressed because of bad decisions they might have made. Other people worry about the future and get anxious that it might not pan out the way they hope or expect. A lot of people get stressed about both sometimes (including myself). The thing is, and I know it's easy to say on paper, but both of those things are futile. The past has already gone and the future isn't yet decided, so why worry?

Faith is simply trusting that things will be okay, whatever happens. Faith is letting go of what isn't important and hoping that the best outcome will occur. Faith isn't naïve or lazy or fatalistic. Faith is being courageous, because it takes guts to believe in something that you can't see right in front of you. Faith is confidence in your own ability. You can grow faith with practice and experience by experimenting.

And of course, it's not only about ourselves, we must have faith in the good of other people; most humans are good after all. My last big expedition took me on a journey of over five thousand miles around the Arabian Peninsula through thirteen countries, including Iraq and Yemen and the Empty Quarter desert. I travelled on foot, camel and dhow, through some of the most hostile desert environments, warzones and waters on earth.

It is a region of which many people are afraid. In our collective imagination we hold so much fear about the Middle East, some of which has become distorted and grown and settled in as prejudice. Before I set off, everyone thought I was crazy. I wanted to pitch the project to the television people, but even with four successful documentaries under my belt, nobody would touch the idea.

'It's too dangerous,' they bleated. 'You'll be killed,' they warned. 'We're not paying your ransom,' they complained.

In fact, the only person who didn't think that I'd disappear and never be seen again was my mum, who simply shrugged and said, 'Have a lovely holiday.'

The point is, my journey was perceived by most as being far too dangerous, because they didn't have faith in the people of that part of the world. I'm not saying that I was being naïve or was looking at the people of the Middle East through rose-tinted spectacles, and I'm not saying that the commissioners were necessarily to blame; the media has been distorting the global view of this part of the world for decades and decades. Don't get me wrong, there are parts of the region that are a stinking mess – from hotbeds of terrorism to a deadly and futile civil war. But equally and far more importantly, large parts are not.

I often get asked if I travel armed. The truth is I'm sometimes obliged to take an armed escort or a policeman when travelling through dangerous areas in places like Syria or Iraq, but I never carry a weapon. If you go somewhere with a gun expecting trouble, you'll probably find it.

It's also worth adding that if you go somewhere with certain expectations, and they don't work out as you planned, then you'll be disappointed and unhappy. You'll come away with a negative perception.

Likewise, if you go somewhere thinking the people are bad, that negativity will rub off and you'll be treated poorly. Only if you take a punt will you ever discover if something is possible. What I have learned from my travels is that wherever you go in the world, people tend to look after each other so long as you go in peace, crush your ego and have a bit of faith in them.

So I set off on my journey around the Middle East hoping for the best. And guess what, the best happened (for the most part). I encountered incredible hospitality and warmth in the least likely places. In the midst of terrible poverty and tragic conflict, I encountered a kindness that I wouldn't have thought possible. I wanted to experience the very best of the Middle East, and I got exactly what I was looking for.

In Iraq, I was taken in by heavily armed militia forces and shown wonderful respect by people who had not so long ago been trying to kill British soldiers. In Yemen, I was looked after by refugees and rebels with nothing of material worth to their name, and in Jordan I spent weeks with the Bedouin in the simplicity of the desert.

In Syria, I was taken in by a lady whose son was a soldier in Assad's army and had been killed by rebel fighters a couple of years previously. She invited me into her bullet-strafed apartment on the front line in Homs and insisted I stay in her son's bedroom, all this in the knowledge that 'her side' was technically the 'enemy'. Never once did I feel endangered by the people who had taken me in, but often I felt incredibly grateful to have met such generous and humble people.

Heroes of our own journey

Travel has long been romanticised and portrayed as a challenge or quest. This is nothing new. Homer's *Odyssey* was the first in a very long line of heroic depictions of travellers going off in search of something. In these stories, our hero must leave behind the world of the familiar and venture into an alien world full of dangers and obstacles. He must overcome them in a series of tests, before meeting his ultimate challenge in whatever form that may come.

Whether you're Saint George needing to slay a dragon, or Frodo Baggins and you must throw the ring into the volcano of Mordor, the hero always has to come face to face with his greatest fear, which of course he does before winning the battle, the girl or whatever the prize may be, before going home a better person to share his spoils with his countrymen.

Think of any Hollywood movie script, Disney film, or the plot of any adventure novel, and they're all basically the same. There's a reason that Paolo Coelho's *The Alchemist* sold 65 million copies. It is utterly predictable with a basic plot that is so formulaic, it is almost as if you know exactly what's coming next. The reason being that you *do know* what is coming next – you've read and watched it a hundred times, even if you didn't realise it. It's called the monomyth, or the twelve phases of the hero's journey, and forms the basis of every story ever told. There isn't enough space to go into the detail here, but Google it, and you'll never watch a film in the same way again.

We have become so enamoured with the concept that travel gives us the opportunity for adventure and glory, that we see ourselves as the hero in our own stories. That's what adventure is all about – living out a fantasy in an alien realm. We get to experience our childhood dreams. Even the most placid beach holiday is an opportunity to go away, try a new cocktail, hear a little of a new language and go home to boast about how tough the airport transfer was.

For those who travel independently and avoid the crowds, this takes us one step further into the realms of adventure. We can play out a role, and be whoever we want to be – especially if we travel alone. There's no one to tell you what to do, and who to be. Strangers never knew the 'you' back home, and it gives anyone the opportunity to start again, or at the very least try out a new way of being, away from the constraints and expectations of home.

But there is a danger in thinking that travelling is the solution to all our problems. Without a focus on empathy and faith, it won't bring us contentment or fulfilment. The point of the stories we tell, to and about ourselves and other people, is that they have a narrative, an overarching meaning that takes us to an end point, having learned something along the way. It's not enough to travel the world; we must do so in a spirit of openness, empathy and good faith, or we are only moving our discontent from one place to another.

Underlying every Hero's Journey narrative is a question of character. Ultimately, the story takes the hero from a situation of relative safety, where their character flaws hold them back but don't get directly tested, to one where they find themselves in mortal peril. At this point, every character faces a choice. They can identify whatever it was that was holding them back, overcome it and save the world or themselves. This is a story we recognise as one with a happy ending – the ancient Greeks called this 'comedy', though the word means something different nowadays.

But there are also tragedies, where the hero (or better, the protagonist) doesn't recognise or admit this character flaw. They persist with whatever vice it was that held them back in the beginning of the story, and this leads inexorably to their downfall.

We can all be the hero of our own journey, but what kind of journey this will be is a choice that lies with us. It ties back to self-knowledge, where we began this particular journey, and requires us to explore the world with faith, empathy, gratitude and ultimately courage. Be the master of your own destiny. You can do this with both confidence and humility, with compassion, bravery and flair all at the same time, without the need for one-upmanship, false modesty or deceit. Ultimately that's the job of the modern explorer: to master oneself and to share one's learnings for the benefit of others.

Share Your Findings

Jacques-Yves Cousteau is the father of marine exploration. His pioneering aquatic discoveries in the 1940s led to the development of modern diving techniques and underwater filming equipment that enabled a new understanding of the world's oceans.

Born in France in 1910, he was a naval officer who assisted the allies in the Second World War, conducting anti-espionage operations against the Italians. After the war, he led teams of marine scientists to push the boundaries of breathing apparatus, so that divers could spend more time underwater and conduct marine surveys and archaeological excavations around the world. He co-invented the aqualung, advanced underwater photography, discovered oil beneath the ocean floor in the Persian Gulf and built undersea stations and small submarines for oceanographic research. He even tried to build an underwater city.

He soon became famous for his explorations and daring discoveries. He was involved in more than 120 television documentaries and contributed to some eighty books, receiving countless medals and accolades in the process.

More importantly, Cousteau was an ardent conservationist and a leader in the movement to protect the world's oceans, and his legacy is a huge body of work that educated an entire generation about the global need to address climate change – inspiring millions of people to take environmentalism seriously.

He used to describe himself as a humble oceanographic technician, but he was no shy boffin. He redefined the notion of exploration in the twentieth century, and demonstrated that an individual could indeed leave a lasting, positive change by sharing his findings with others.

Jacques Cousteau summed up his solid self-confidence and aversion to false modesty rather well: 'When one man, for whatever reason, has the opportunity to lead an extraordinary life, he has no right to keep it to himself.'

Be grateful to be alive

In 2015, I was in Nepal as part of a journey to walk the length of the Himalayas, some 2,000 miles from the Wakhan corridor in Afghanistan to the mountainous kingdom of Bhutan. I was with my guide, Binod, and my brother, Pete, who had taken some valuable holiday from his job in the UK and flown out to join me for a few weeks. It had been a tricky time; the monsoon had struck with a vengeance making many of the roads impassable, or at the very least slippery and dangerous. What is more, the Maoists were on strike, which meant that the majority of restaurants and tea houses, where we might be able to eat or sleep, were closed.

We were trekking in the rain, day in day out, and it was pretty demoralising. One evening we made it to a tiny remote village high in the mountains and Binod set about trying to find us a place to stay. It was getting late and we were soaked through, but no one was able to house us and there was no place for us to camp. Due to the strike, all driving was banned, but he managed to persuade a taxi to take us at dusk up to the nearby town of Musikot, where we would be able to find somewhere safe and dry to sleep.

It wasn't the most reassuring thought; I knew the roads in the mountains were lethal and many of the drivers seemed to be verging on suicidal. I had witnessed accidents and the wet weather was going to make it even more risky, but it was getting dark and it was not safe to walk any further in the darkness. I tried to ignore my nerves as we clambered into the battered 4x4, and I noticed that like most cars in that part of the world there were no seatbelts.

So we drove into the pitch black, spluttering and winding up the narrow roads with thick forest on either side of us. We had not gone far and I was beginning to relax, lulled into a false sense of security as we crested the top of the mountain pass. Suddenly I felt the car lurch and then fly forwards. I immediately recognised that the brake cable had snapped.

The driver was panicked and hadn't managed to get down through the gears quickly enough to slow us down, or to steer us into the cliff. I remember yelling at him to hit the wall, begging him to drive into the cliff. While that might sound mad and totally counterintuitive, in the mountains it's not. The alternative, on the other side of the road, was much, much worse. I knew that at some stage we'd have to come off the road, and that's exactly what happened.

As we gathered speed, I steeled myself.

'Hit the wall!' I shouted, 'Turn the wheel right and hit the wall!'

I begged in vain as I felt the car lurch out of control. I gripped at the dashboard and instantly froze, not daring to even look at the man supposedly in control to my right. He was too busy battling with the gears to see the logic in my cry. We were now rocketing straight down the road and all I could see were flashes of rock and low-lying branches as the headlights flashed into what lay before us. If only he had turned into the wall when I first shouted, we'd have crashed, but at least we'd all have survived unscathed.

That was no longer an option. We'd missed the window of opportunity and as we flew aimlessly in the darkness of the night, my body crippled under the terror of what could only happen next. The car jerked sideways as the front tyre clipped an invisible drainage ditch, the impact causing the vehicle to veer left until it was almost on two wheels, the momentum careering us faster and closer to the inevitable; plunging straight off the edge of a vertical cliff, hundreds of feet into a jungle ravine in the dead of night.

The next thing I can clearly recall was thinking the word 'No'. It echoed through my body. I didn't want it to end like this and certainly not with my brother and Binod in the car too. They couldn't die.

We went over, we went down, and somehow, we lived. I remember coming to an abrupt stop; I was upside down and with searing pain coursing through my upper body. That was reassuring at least, because it meant I was alive. As I realised this, my survival instinct kicked in. I yelled for Pete and Binod, desperately calling out their names to see if they were there. I remember willing Pete to respond; I was bellowing for help into the darkness. I couldn't focus. I was paralysed by dread at the thought of my brother being gone, and I wanted the nightmare to be over. I used all my energy to yell into the night.

'Pete, are you alive?'

'Lev!' Pete shouted. 'I'm here.'

I've never been so glad in all my life to hear someone's voice. The relief gave me renewed energy. Miraculously, Pete was in better shape than I was, with the fewest injuries, and he was able to find a torch in the car and he started flashing it into the dark. Suddenly Pete spotted lights, there was a chance of hope.

I was hallucinating by this point, determined that we weren't really alive, and that this was hell. I tried to convince Pete that there was no use, they were too far away and they would never find us; we were going to stay in that valley forever. Thankfully, Pete had the presence of mind to ignore my pain-induced ramblings and he screamed and screamed in the direction of their lights. They must have heard the crash or my yells of pain, or seen the lights. Either way, after a while – I couldn't tell you how long – dozens of villagers appeared.

Only then did I believe that we had been saved. I have a blurry recollection of being scooped up on a very makeshift stretcher and dragged out of the jungle. I remember thinking about my Paras training, where we'd have to learn how to carry the weight of an injured man off a battlefield, and here were these diminutive Nepali men, doing it for real – through dense jungle and crossing streams up to their waists. It was agony, bumping and crashing through the thick forest, and they dropped me more than once. But it didn't matter. We had been rescued.

I came round a day or so later, hazy from the painkillers, to find myself in a filthy, dirty, ramshackle clinic high in the mountains. We were nursed here for three days, because the monsoon and the heavy rains meant that it was impossible to be evacuated either by road, due to landslides, or by helicopter, due to the low visibility. It was only a week later that I was able to get surgery to fix my broken bones and fully comprehend what had happened.

Of course, it was a disaster. I had broken my arm badly and was unlikely ever to get full and proper use of my shoulder again; the expedition had ground to a halt and my guide Binod, as well as my brother, had both sustained physical and emotional injuries that might have untold consequences. There were times as I lay in agony, with my arm twisted and shattered, when I contemplated giving it all up, the world of exploring and expeditions. I questioned my own sanity; what on earth was I doing taking these risks anyway?

I would later learn that we had rolled and bounced almost 200 feet down the side of a cliff into the valley below. I was shocked into self doubt and mortal fear. What was the point of it all? I thought about packing it all in and getting a regular 9–5 job. It was that serious, but then I was reminded of a quote by J.R.R. Tolkien: 'Faithless is he that says farewell when the road darkens.'

I knew deep down that I couldn't just give up, no matter how hard it was to contemplate carrying on. I counted my blessings instead; I knew I was extremely fortunate to be alive. Looking back now, this was the clearest moment in my life where having a sense of hope for the future was essential to my inward happiness. That could only be engendered by positivity towards myself and my own worth. We can choose our own destiny, in effect, if only we have the courage to take a leap of faith. Even in the most trying of circumstances, hope can prevail, and that for me has been one of the most astounding discoveries on any of my travels.

So I persevered (albeit accompanied by a forgivable and deep-seated fear of travelling in the back of cars on mountain roads, which I still cannot shake) and I journeyed on. After a few weeks off to let my arm heal, I returned to Nepal to continue the expedition.

Skip forward three and a half years to 2019 and I am back in Nepal. I have arranged for Pete, Binod and I to return to the site of the crash. We go by helicopter this time, instead of driving, and as we soar over the lush foothills of the Himalayas, I start to contemplate properly what is happening.

What is the point of this trip? What am I trying to achieve? I know it was hard for both Pete and Binod to get to grips with the accident. I can only imagine how terrifying it must have been for Pete, as the one who was the least injured; he must have felt a huge responsibility to get us out of there. He then had to watch us both drugged up and out of it in that clinic, while he had to go through the stress of trying to get us airlifted out.

What about me? I have had no nightmares or flashbacks about the accident, which might be considered a bit odd. But trauma can present itself in strange ways. I think that night's experience came for me in other ways. I have always been restless, but for a long time after I became even more so. I would go off on long expeditions, spending more and more time away from friends and family, and burying my head in my work. I was holding onto a lot of guilt, I resented myself for putting my brother and friends in mortal danger and of course my poor family who I knew must worry about me every day.

As we swoop down to land on the remote, dusty airstrip, I try to stop overthinking it all. We are nearly here now and I want to take in the experience for what it is; I will never come back here again. We clamber out of the helicopter into the dust and set off towards the crash site, trailed by a troupe of children and adults curious to see what we're up to. We traipse along the road, in a rather sombre and reflective silence, until on rounding a corner, we see the car.

It is mangled, completely destroyed, sitting there rotting on the side of the road, unceremonious. We are quiet, filled with disbelief that we are still here. We stare at the remains of the crumbling 4x4 that nearly cut our lives short, but also saved them. Our silence is punctuated by a group of nearby villagers, as quite a crowd has gathered, clearly wondering why we are paying homage to a rusting hulk of metal. Suddenly, looking around at the curious, friendly faces, I feel a warm rush of gratitude. Any negativity I feel about the whole affair is dissipated.

Standing there that day, I realised there was no use pointing fingers of blame at anyone – even myself. I felt a huge amount of gratitude to the people who had saved my life. How had they found us and why had they come out to look for us? My mind raced with alternate endings. What if they hadn't heard us – or if they had heard our cries and ignored us?

I wondered for a second what would happen in other parts of the world. Here in Nepal I had been looked after as a brother on many occasions, fed to bursting and given a place to sleep by perfect strangers. Binod himself had done that for me when I first met him as a young traveller all those years before, and looking into his eyes now, I saw nothing but humble compassion.

I was grateful too for everything that I've had the chance to do since then. I think that accident changed me for the better in so many ways. It gave me a newfound respect for my life and a deep sense of gratitude that I had never felt before; it also gave me a fortitude that I'd not had either. After the accident I could have given up, but I would never have forgiven myself. I had two choices; I could admit defeat and go home, or I could see it as a blessing in disguise and crack on. That was entirely up to me. Everything in life is a choice, including how we react to the things that happen to us. That even goes for our emotions: we can choose to be happy as much as we can choose to be sad.

Standing there that day, staring at the wreck of the car, I also felt a huge amount of gratitude towards my brother, not only for saving my life, but for his friendship. I am extremely lucky to be alive. It's a funny thing to say that and for it to be factually correct, when usually it's such a throwaway line, but I seriously am. It took me falling off a 200-foot cliff to really appreciate what that truly means.

Not all of our challenges will involve climbing mountains, or crossing deserts. A lot of our greatest battles are with our mind, and by returning to the crash with the people I loved, I felt the same flood of emotions as I had at the end of some of my most arduous explorations.

More often that not our own Everest is resting within us, simply waiting to be conquered.

Nims Purja

Nirmal Purja, or Nims, as he goes by, grew up in a small village in the lowlands of Nepal. At eighteen, he passed the gruelling selection process to join the Gurkhas in the British Army. Nims went on to serve in the Special Forces, but in 2019, after sixteen years of service, he dedicated himself to a new goal – a world record climbing attempt, which he named Project Possible.

Nims's mission was to summit the fourteen highest mountains on the planet, all of which are over 8,000 metres tall in the Himalayas. The last person to succeed took almost eight years, but Nims planned to pull it off in just seven months. Nims only came to climbing in his late twenties, when he got fed up with telling people that he had never seen Everest – his home village was miles away from the region and no more than 500 metres above sea level.

So, determined to clap eyes on Everest, he decided to trek to base camp and it was while winding his way up the narrow pathways that he saw the distinctive shape of a mountain called Ama Dablam. At 6,812 metres, it is certainly no beginner's climb, but Nims persuaded his guide to let him do it – learning how to use crampons in only a few days – and the pair successfully reached the peak.

Project Possible was another beast altogether. Above 8,000 metres, in what is known as the Death Zone, your body starts to shut down. With no oxygen you are slowed to a snail's pace, expending no spare energy and focusing overwhelmingly on keeping the extremities warm. It takes incredible mental fortitude to keep on going up, instead of turning around and going down. Physical risks aside, expeditions like this are phenomenally pricey; faced by a shortfall of funding, Nims left his job, forgoing his army pension, and remortgaged his house.

Nims achieved his goal of climbing the world's fourteen tallest peaks in the record time of six months and six days. In doing so, he bagged a further six world records along the way. If that wasn't incredible enough, Nims risked not only his record attempts but also his own life on multiple occasions, to aid in the rescue missions of other climbers. That is the sign of a real hero. Even with all those things at stake, he put the lives of others ahead of his own goals.

Nims's attitude is remarkable. His positive mindset combined with a steely determination allowed him to achieve the impossible, and yet despite it all he remained humble. Nims is a man who embodies the Art of Exploration; keeping the destination in sight, but always remembering to enjoy the journey.

Happiness or fulfilment?

Learning and understanding what makes us happy is no easy feat. If we can observe when we are expressing ourselves most purely, contentedly, and being honest, then we are a long way towards satisfaction. But there is no magic recipe for contentment, and it is hard to assess, because hindsight can alter our perception significantly. On my walking expeditions, I have covered somewhere in the region of ten thousand miles on foot. I have often looked back and pondered how much of that time spent plodding I was actually happy.

Even now it's hard to judge, because we tend to forget the bad bits, and we are only reminded on our Facebook memories of the moments we shared those years ago, which were inevitably the good ones. I do recall that there were plenty of low times, and I've still got the scars to prove it, but they have given me food for thought as to what keeps me going during the toughest times. I now realise how much I appreciate companionship, and it is important to me that I share these journeys with my friends and loved ones. As Mark Twain said, 'to get the full value of joy you must have somebody to divide it with.'

I have also learnt the crucial difference between happiness and fulfilment: one is fleeting, but the other is long-lasting. Happiness is euphoric and makes your heart race, while fulfilment is serene and makes your heart sing. Casting off expectation is the key to fulfilment; it is about peace, not quick hits or immediate fixes. I would say that happiness is a byproduct of fulfilment.

One-off events or incidents can make you feel happy, but

fulfilment is deeper; it's about finding your purpose and a sense of meaning. I am unlikely to come out of a meeting and say, that went well, I feel really fulfilled; but I might get into bed at night thinking, that was a good day, I feel really fulfilled.

Happiness is also often artificial, and fulfilment is natural. Fulfilment is about achievement and about satisfaction; something bigger that you can achieve through time and effort, whilst happiness is something that you might feel along the way.

Think of happiness as short, sharp bursts of energy and excitement to keep you going, dopamine and endorphins flooding through your body; but this is temporary. We can seldom call upon the feeling of happiness from something that happened to us a few years ago. It is intense, but it fades.

Fulfilment on the other hand is far more enduring, far deeper and more all-encompassing. Happiness is a mood, while fulfilment is a life goal. It requires a lot more dedication, too. In my experience, it requires two things if you want to achieve it.

Purpose and compassion. Firstly, find out what it is that drives you, discover your higher calling and believe in it. Stick at it through the tough times. It won't always be easy, or fun, but it will be worth it. By following your dreams there is a good chance you will discover true fulfilment, because you are being your true self.

The second requirement is compassion, for it is only by letting go of our egos and giving of ourselves that we can truly feel fulfilment. Compassion is born out of empathy and it is the human emotion that instils peace of mind, and helps us in sticking up for and protecting those in suffering. A compassionate person is able to use their empathy to be accepting of themselves, and accepting of others.

The Dalai Lama says that the more compassionate our mind, the better it will function. The opposite of this is letting in fear and anger, and as we all know, that never leads to anything good. And he is right, compassion acts as a buffer against stress and anxiety; it gives us confidence in our own abilities and imbues us with an inner strength, as well as keeping us calm. As His Holiness once said to me, in the glistening courtyard of his Himalayan retreat, 'Helping others is the best way to help yourself, the best way to promote your own happiness. It is you, yourself, who will receive the benefit.'

Who knows, maybe there is no such thing as a truly selfless act, and that all charity is done to serve yourself in some way. Perhaps this is true, but does it really matter if you are doing a kind deed? Whatever the motivation, true exploration is to know that you have had a positive impact on the world around you; that you have done more good than harm. When you adopt this attitude and accept that true discovery is inside us, then the explorer within you will never tire, never fade and never give up.

The Art of Exploration is simple really. It is in realising that world is a far better place than others would have you believe. So go out and see it with your own eyes and rediscover your own childlike sense of wonder, because it is in that joy that you will find true happiness. While you are on the path to discovery, tread lightly and leave the world a better place than you found it. Inspire and help others to achieve their dreams. Judge people less and focus on improving yourself more.

We can all do better. Live a life that you are proud of and remember that it is never too late to start over again. Face your fears and have the courage to choose the hard path; by embracing the unknown, the universe rewards you with hidden treasures. Magic is everywhere, if only you will open your eyes to see it.

The essence of the art of exploration is already inside you.

Explore, Dream, Discover.

The author in Afghanistan, 2008

Acknowledgements

As you will have gathered if you have read this book, I have been fortunate to receive the guidance and wisdom of a great many people along the way, and it would be rather difficult to thank each and every person for their help. I hope that telling their stories will be sufficient recompense.

Yet I must do my best to pay my thanks to those that have taken this journey with me, some over the course of many years, some only briefly but who, for whatever reason made a lasting impact. There are no doubt many that I have overlooked, or whose stories or advice were not appropriate for this particular book, but have inspired me in other ways. I promise if you remind me of my oversight I will buy you a drink!

This book is the result of over three years writing, in between journeys and other projects, and was impacted significantly by the events of COVID-19. But really it's an idea that goes back much further, bringing together some of the memories of the first half of my life. I would like to thank my publishers Hodder & Stoughton for their ongoing support, in particular my editor Rupert Lancaster as well as Cameron Myers, Caitriona Horne, Rebecca Mundy and Barry Johnston.

As ever, I owe the book to my agent Jo Cantello at Wolfsong Media, and my eternal gratitude to all those who helped with the editing process, read the drafts and provided a watchful eye in keeping me on track. Special thanks to Charlotte Tottenham, Kate Harrison, Alex Eslick, Daniel McEvoy, Emma Challinor, Stephanie Lennard, Charlotte Mackintosh and Geraint Jones.

And in no particular order appreciation must go to the following for their words of wisdom, sage advice and companionship along the way: His Holiness the Dalai Lama, Sir Michael Palin, Sir Ran Fiennes, Bruce Parry, Ed Stafford, Helen Sharman, Paul Adams, all my teachers at St Peters CofE School and Painsley R.C. College, Dr.

Ross Balzaretti and all my professors at the University of Nottingham, my instructors at RMAS, brother Officers and soldiers of the Parachute Regiment, Ceci Alonzo, Ash Bhardwaj, Dave Luke, Tom Bodkin, Chris Mahoney, Neil Bonner, John Copeland, Alberto Caceres, Will Charlton, Vivien Godfrey, Will Storr, Sophy Roberts, Henny Hardy, Tom Shore, Tracey Saunders, Alex Bescoby, Kate Page, Max Clark, my brother Pete and my wonderful parents for their eternal support and inspiration

The Art of Exploration

Also by Levison Wood

Walking the Nile
Walking the Himalayas
Walking the Americas
Eastern Horizons
Arabia
Incredible Journeys
The Last Giants
Encounters